THE SOLDIERS OF
SAN JACINTO

By Johnnie Belle McDonald

Edited by J. Patrick McCord & Michelle M. Haas

Copano Bay Press
2008

*Originally published in 1922 at the University of Texas
under the same title as J. B. McDonald's M.A. thesis.*

ISBN: 978-0-9822467-2-6

Contents

Publisher's Note

The Battle of San Jacinto was an event that not only marked the dawn of a short-lived Republic, but was one that also changed the political, cultural and cartographic landscape of the North American continent. Those eighteen intense minutes literally put Texas on the map.

There will forever be conflicting data about the total number of men who fought in the Battle since the regiments were not afforded the luxury of drilling together and the men scarcely stayed put long enough for formal roll call. The total numbers stated here differ from other counts, though not widely. There are names present here that are not present in other writing on the subject and vice versa. In publishing this volume, we hope to offer Texans another resource to complement those already published on the men who fought in the Battle of San Jacinto, and perhaps to offer alternative data for some of the names found here. The reader should approach this information with the view that it was dredged out of land records in 1922 and can only be as accurate and complete as were the original records as of that year. The spelling of the vast majority of names was left as Miss McDonald originally recorded them, as was most of the pertinent data. Organizational and grammatical changes were made to render the book more easily read.

This text has a higher purpose beyond that of being a useful research document. The notion that the Texas Revolution and its final battle were fought primarily by American mercenaries, filibusters or bored U.S. soldiers looking for action is still prevalent today. Miss McDonald's research illustrates quite a different view of the men who fought in the Battle of San Jacinto, a view that demonstrates that roughly half of the men who fought it were in Texas prior to her formal Declaration of Independence and that most of the fighters made Texas their home after it was over. The data presented here makes it abundantly clear that Texians (men born in Mexican Texas and immigrant men who sought to make a life here and live as Texans) fought and won Texas independence from Mexico.

Michelle M. Haas, Managing Editor
Seven Palms - Rockport, Texas

4:00 PM, April 21, 1836

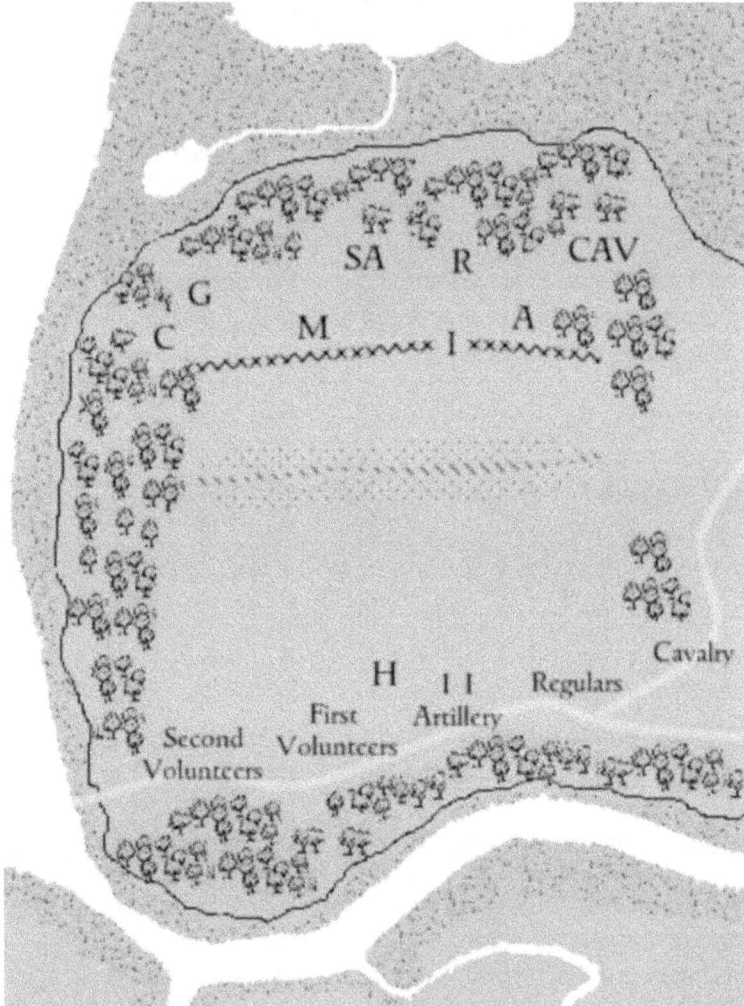

A - Aldama Battalion   R - Mexican reserves
C - Cos' reinforcements   SA - Santa Anna
G - Guerrero Battalion   CAV - Mexican Cavalry
M - Matamoros Battalion   I - Cannon

H - General Houston

PART ONE:

THE BATTLE OF SAN JACINTO
IN 900 WORDS
BY MARK PUSATERI

4:00 PM, April 21, 1836 — General Houston, atop a dappled gray, gives the order, "Trail arms! Foward!" Some 900 men, unwashed, underfed, caked with mud and dressed in rags, begin a long walk through knee-high grass. They have been pushed to the edge, run from their homes, their crops and houses burned. They don't know whether their families have found safety. They've lost kin and good friends at the Alamo and at Goliad. They want a fight and are about to get it.

At the far left of this parade line is the Second Regiment of Volunteers, 330 men under Colonel Sherman. To their right, at the center of the Texian force, is Colonel Burleson's First Volunteers, 386 men strong. Next are the 32 men of Colonel Hockley's Artillery Corps. They man two iron canon, six-pounders called the Twin Sisters, gifts from the people of Cincinnati. To the right of the artillery are 92 men of the Regular Army under Lt. Colonel Millard. At the extreme right is the Cavalry, 62 mounted men commanded by Colonel Mirabeau Lamar, just yesterday a private. All advance in perfect silence.

4:30 PM

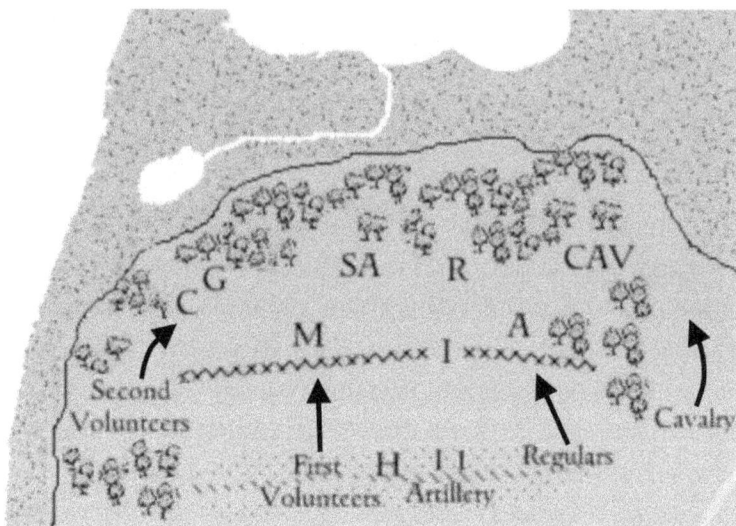

4:30 PM — The Second Volunteers under Sherman, having traveled swiftly through the oaks on the Texian left, fire on the surprised men of General Cos' command. The Battle of San Jacinto has begun. The Mexican forces return fire, but they are soon on the run. Sherman, leading the pursuit, is the first to shout, "Remember the Alamo! Remember Goliad!" The main body of the Texian forces crest a slight rise. They are 200 yards from the Mexican breastworks, a four foot barricade of cut brush, saddles and baggage. Houston, riding thirty yards in front of the First Regiment, orders, "To the charge! To the charge!" Musicians strike up a bawdy march on fife, drum and fiddle. General Castrillón directs his canon fire on Lamar's advancing cavalry. The Twin Sisters, loaded with cut-up horse shoes, hail hot metal at the alarmed Mexican troops. A small force advances on the Texian artillery, but is repulsed.

4:35 PM

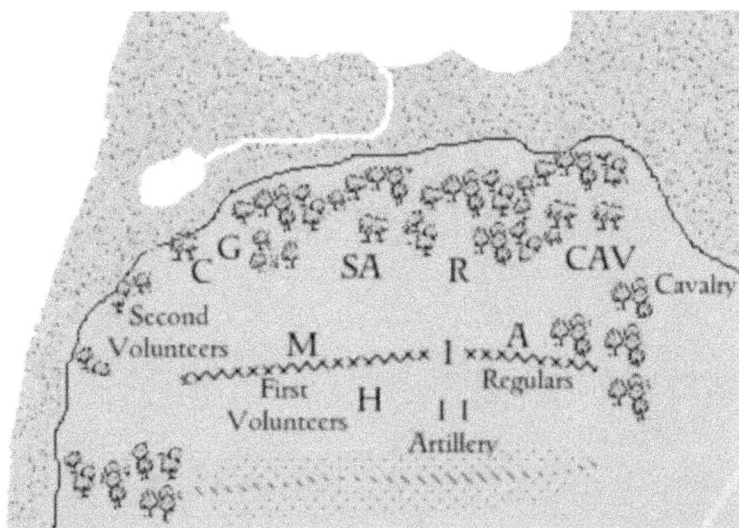

4:35 PM — Havoc reigns on the enemy left as the Texian Cavalry attack their stunned counterparts with slashing sabers. Burleson's First Volunteers are upon the breast-works engaging Matamoros Battalion. To their right Texian regulars assault Aldama Battalion with equal ferocity. Stampeding behind the lines, riderless Mexican horses bring terror to the breastworks defenders, who now believe they are being attacked from the rear. The Second Volunteers drive Cos' panicked men rearward into Colonel Almonte's Guerrero Regiment, pushing them all nearly two hundred yards.

4:40 PM

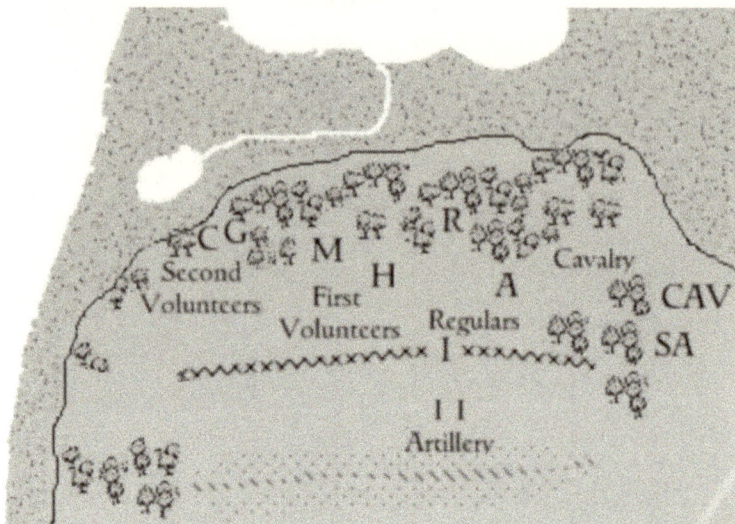

4:40 PM — Almonte attempts to rally any men who can still be commanded, but it is too late. Matamoros and Aldama Battalions turn from defense of the breastworks in wild retreat. The First Volunteers and Texian Regulars are over the breastworks, pursuing with savage intent. The resistance at the Mexican canon position is overcome and the gun seized. Any Mexican cavalry able to mount up flee toward Harrisburg, Santa Anna among them.

4:45 PM — Sherman's Second Volunteers chase Cos and Almonte's men into a small bayou to the Texian left. The First Volunteers force Matamoros Battalion into the marsh at the rear of the Mexican position and into Peggy's Lake. Some try to surrender, pleading for their lives, crying, "Me no Alamo! Me no la Bahia!" There is no mercy. Many Texians fire only once and don't waste time to reload. They turn their rifles around and swing them as war clubs, breaking many off at the breach in the act of shattering a skull. The air is filled with the acrid smell of gun powder and the stench of feces as dying men void their bowels.

4:48 PM

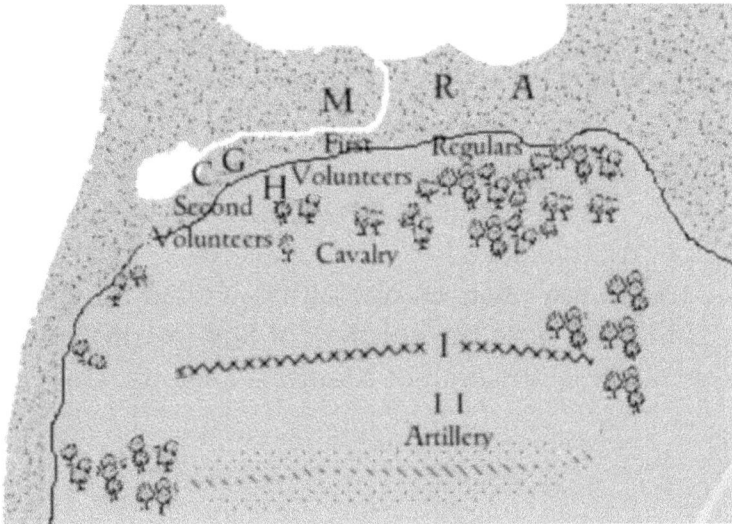

4:48 PM — The Battle of San Jacinto is over, but not the killing. Behind the Texians are the enemy dead. To their front, in marsh, lake and bayou, those Santanistas still living try in vain to escape or plead for their lives. The Texians calmly, but briskly reload, time and again. Each shot means the end for another of Soldado.

Sundown — A guard is set on the Mexican camp to keep the men from looting. The spoils are to be divided among them as war booty. Mexican soldiers who escaped the slaughter are being rounded up and marched to the oak stand on Buffalo Bayou from which the Texians set out barely two and a half hours ago. They will be held in a pen made of split logs, rope and anything else that lends itself to the job. The Texians wander back to camp, singly and in small groups. Some talk about deeds of the day, others sing songs, laugh and trade cheers across the prairie. Still others just walk, their thoughts their own until the end of their days.

If the Alamo is called Texas' Thermopylae, then San Jacinto is her Agincourt. Of the Texian forces there are but seven killed. Twenty-nine are wounded, including General Houston, his ankle shattered by a copper ball from an *escopeta*. Of those wounded four will die. The Mexican dead number 630. The prisoners tally 730, of which 208 are wounded. The events of this day will mean perpetual freedom for Texas, as a republic for now, and in ten years as one of the United States. History will show that the soldiers of San Jacinto have set the keystone in the arch of Manifest Destiny.

PART TWO
THE OBJECTIVE

I have made a careful study of the available records to determine primarily how long the soldiers of San Jacinto had been in Texas at the time of the Battle, and also their place of origin. This study was necessary, because it has often been asserted in works of apparent responsibility that the Battle of San Jacinto was mainly won by Americans visiting Texas to fight the Mexicans. Santa Anna said in his *Manifesto*, published in 1837:

> It is notorious that the soldiers of Travis in the Alamo, those of Fannin at Coleto, the rifleman of Doctor Grant, and Houston himself, and the troops of San Jacinto, with very few exceptions, came from New Orleans and other points of the neighboring republic, exclusively to support the rebellion in Texas, having had no previous relation with the Colonists or their enterprises.[1]

Doctor Herman von Holst, in his *Constitutional and Political History of the United States*, implies that the majority of the San Jacinto soldiers came directly from the United States purposely to fight the Battle. He says:

> There was no longer any time for delay but yet the decision was given by citizens of the United States. It was given on the 21st of April at San Jacinto. Santa Anna, who led the Battle 'personally,' is said to have had one thousand five hundred men, while the 'Texas Army' under Houston amounted to only eight hundred men, of whom it is said not more than fifty were citizens of Texas.[2]

Von Holst quotes the following from *Niles Register*, LXIV, 174:

> Wise, of Virginia, certainly a witness not to be thought too lightly of, said, in 1842, in the house of representatives: 'It was they (the people of the great valley of the Mississippi)

that conquered Santa Anna at San Jacinto; and three fourths of them, after winning that glorious field, has peaceably returned to their homes' (that is, to the United States.)[3]

As soon as the news reached Washington that Santa Anna was marching to Texas with a large army, President Jackson ordered General Edmund Pendleton Gaines, the commanding officer in the South, to proceed to "some proper position near the western frontier of the State of Louisiana," and to see that neither of the contending parties crossed the boundary into the United States, and to keep the Indians from making hostile incursions into Texas from the United States.[4] Von Holst, in *The Constitutional and Political History of the United States*, indicates that General Gaines' sympathies were with Texas, and that a number of his soldiers joined them in their fight for independence. In his discussion of the question he said:

> The Texan Army which wished to grant its powerful protection to the imperiled Union against the Mexicans and Indians, consisted, indeed, no longer of citizens of the United States, but in part of troops of the Union. Gaines' soldiers kept on their uniforms, but preferred to go to the Texans, where there was something to do; and when the Union officers for shame's sake demanded their deserters back, their Texan colleagues answered with a 'pitying' shrug of the shoulders.[5]

McMaster in discussing the same subject in *A History of the People of the United States* says:

> Later in the year when Gaines sent an officer into Texas to claim deserters back from his army, they were found two hundred in number enlisted in the service of the young Republic and still wearing the uniform of the United States.[6]

These statements are typical, and I have used them to show the tenacity of the traditions. The result of this study shows that a large percentage of the Texan Army at San Jacinto was composed of men already settled in Texas or who afterward became permanent residents. Judge Seth Shepard was substantially correct when he wrote concerning the Texan soldiers:

It is a great mistake and a gross injustice to their memories to suppose that these men were American 'filibusters' or revolutionists by nature, and that their purpose in coming to Texas was ultimately to sever it from Mexico. They were, in the main, men of peace and lovers of good order. They came in good faith to make homes and to perform all the obligations of their compact, and they did perform them as long as the Central power permitted them to do so in peace and safety.[7]

Most of the material for this study is in the manuscript records of the General Land Office. Grateful appreciation is due Honorable J. T. Robison, Commissioner, for the privilege of using the records and to the following members of his staff for valuable assistance and suggestions: to Mr. Charles M. Callaway, who has been connected with the office since May 8, 1870; to Captain W. C. Walsh, who has been in the office since 1856; to Mr. J. H. Walker, Chief Clerk; and to Miss Priscilla H. Buckley, who is in charge of the Spanish archives.

I began my work by using the muster rolls to get a list of the soldiers. Captain Walsh gave me a history of the muster rolls. He said that these copies have been in the office as long as he has been there, that is, since 1856. He does not know exactly when the copies were compiled, but believes that they were made shortly after 1836. They were prepared by the adjutant general from original material in his office. The adjutant general made two copies and gave one to the General Land Office. In 1855 the adjutant general's office was burned. His copy and the original records on which it was based were destroyed. In 1921, Mr. Robison had a typewritten copy made of the old rolls, which is much easier to follow than the old one.

I checked the list of San Jacinto soldiers found in the Land Office rolls with two other lists. One of these is in John Henry Brown, *History of Texas*, II, pages 31-39; the other is in a pamphlet copy of General Houston's official report, published in 1836. The Land Office copy contains all the names given in both of the other lists, and has a number of additional names. Names appear in the three lists with different spellings, and by checking with the original papers in the Land Office, I have been able to determine and give the correct form.

The information which I sought to obtain for each soldier was:

1. Name and company.

2. Birth, date and place.

3. Death, date and place.

4. Date of arrival in Texas.

5. Date of joining army.

6. Family in Texas: (a) sons, (b) daughters, (c) slaves.

7. Other military service with dates.

8. Land acquired from State, with dates, amount and location, for what service granted, whether located by grantee, heirs, or assignee.

9. Migration of self and family before coming to Texas, with dates.

10. Route and method of travel to Texas.

11. Miscellaneous (for example, wealth and occupation).

Having arranged the names in alphabetical order, each on a separate sheet, I consulted first the lists of applicants for land before 1836. Books A and B in the Spanish Records Department of the Land Office contain the names of applicants in Austin's colonies, and there is a separate volume for the respective colonies of Robertson, Milam and Wavell. In these, information concerning each applicant is given in tabulated form. In some instances I found considerable information about the applicants, but in other cases there is little more than the name.

I next used a large volume called *An Index to Mexican and Spanish Titles*. This book contains all the names of the colonists who received land from the Mexican Government before 1836, and after each name there is a tabular statement showing the date of the grant, the amount granted, and its location. This index refers to sixty-eight volumes of original application and titles in the Colonies of Austin; Austin and Williams; De Leon; Zavala; DeWitt; Grant, Durst and Williams; Power and Hewitson; Milam; McMullen and McGloin; Vehlein; Burnet; and Robertson; and to volumes of titles issued by special commissioners G. W. Smyth and Charles S. Taylor in East Texas. Reference to these volumes yields little additional information. The applications are written on officially sealed and stamped paper of the Government of Coahuila y Texas. Neither the application nor the title usually supplies the date of the colonist's arrival or the state from which he came.[8]

Certain collateral documents, however, add valuable information. Both Federal and State colonization laws required colonists to be of good moral character, and the prospective colonist usually filed with his application for land a character certificate and testimonials. Many of them were written by judges, notaries and other officials of the United States, and were brought by the immigrant. Others were written after his arrival by friends already in Texas.

In addition on these records, I carefully examined two volumes containing the titles granted by the different land offices of the Republic between 1836 and 1838, prior to the creation of the General Land Office. The names in these are in alphabetical order and give the date of arrival in Texas, with the amount of land granted, and the board which granted it.

Applications received after the organization of the General Land Office were not recorded in books, but the original applications and related documents were enclosed in an envelope or folder, and these are now preserved in hundreds of filing cases in the Land Office. I was not permitted to handle these original documents, because the laws regulating the administration of the Land Office require such work to be done under the rather minute supervision of an office official, and Commissioner Robison was naturally unable to detail a clerk for such service, which would probably have occupied several weeks. It is quite certain that from this source additional biographical items could be gleaned concerning the 827 men about whom I found something elsewhere, not to mention the 48 of whom I was unable to learn anything, except that they fought in the Battle. I found some information regarding the San Jacinto soldiers scattered at random in the muster roll.

Besides this material gathered at the Land Office, I found some valuable information in Burlage and Hollingsworth's *Abstracts of Land Titles*, published in 1859. This volume includes, among much other material, a number of classified lists of grantees who were recommended for and received grants of land. The lists which are useful for this study are "the first class," "the second class," the list of those who received "donations," and the list of those who received "bounties." The first class is made up of those to whom certificates were issued who were residing in Texas on or before the date of the Declaration of Independence.[9]

The second class is composed of volunteers who arrived in Texas after March 2, and before August 1, 1836, and who either received honorable discharge or died. These received the same quantity of land as those of the first class. The list of those who received donations includes respectively those "who participated in the Battle of San Jacinto, the Siege of Bexar, in the action of the 19th of March, 1836, under the command of Cols. Fannin and Ward, and those who fell at the Alamo, under the command of Bowie and Travis." The amount granted to these men or their heirs was 640 acres.[10]

Bounty warrants were issued for military service, either during the Revolution or the period of the Republic — "320 acres for 3 months' service, 640 acres for 6 months' service, 1280 for 12 months' service or upwards, or for entering the army for the duration of the war." An additional 640 acres was allowed if the soldier died or was killed in the service. Some grants of 320 and 240 acres were issued for special service.[11]

I found some information in Brown's *History of Texas*, Volume II, Thrall's *Pictorial History of Texas*, T. J. Green's *Mier Expedition*, and the following articles: James Washington Winters, "An Account of the Battle of San Jacinto", in the *Quarterly of the Texas State Historical Association*, VI, 139-145: James E. Winston, "Kentucky and the Independence of Texas", in the *Southwestern Historical Quarterly*, XVI, 27-63; James E. Winston, "Virginia and the Independence of Texas", in the *Southwestern Historical Quarterly*, XVI, 277-284; Robinson, "Recollections of Joel W. Robinson", in the *Quarterly of the Texas State Historical Association*, VI, 241-247.

According to the sources that I have utilized, there were 875 soldiers, including officers and privates, in the Battle of San Jacinto,[12] of those I have been able to fix the date of the arrival of 465 and the approximate, or "on or before" date of 242 others. Of the 875 San Jacinto soldiers, 780 obtained land grants, leaving, so far as the records show, 97 who received nothing. Of the 780 who received land, 171 got titles before the Battle and 775 after the Battle, while 166 acquired land both before and after the Battle.

Of the earlier migrations of the families of San Jacinto veterans, little was found in the Land Office records, and the state from which the soldiers themselves immediately emigrated to Texas could be found for only 270. Of these Tennessee furnished 56, Alabama furnished 34, Kentucky furnished 26, Virginia furnished 17, Louisiana furnished 14, Mississippi furnished 10, Arkansas furnished 10, Georgia furnished 8, Missouri furnished 9, North Carolina furnished 5, South Carolina furnished 5, New York furnished 15, Ohio furnished 7, Maine furnished 5, Pennsylvania furnished 8, Massachusetts furnished 5, Illinois furnished 4, Maryland furnished 3, Michigan furnished 1, Indiana furnished 2, England furnished 2, France furnished 1, Ireland furnished 4, Germany furnished 2, Mexicans furnished 15 and the state of origin for two soldiers from New England and Europe could not be determined.

The occupation of only 64 soldiers was specifically stated, as follows: 44 farmers, 6 lawyers, 2 merchants, 2 teachers, 2 bootmakers, 1 clerk, 1 blacksmith, 1 printer, and 5 doctors. The great majority of the others were, of course, farmers.

This study is not as conclusive as could be wished. The arrival of 170 known participants in the Battle of San Jacinto could not be dated even approximately. It is to be hoped that further study of original files in the Land Office may yield additional information about them. It is possible, of course, the deserters from Gaines' army and volunteers from the United States fought and then returned to the United States without asking for compensation. This does not seem probable, but this study has not disclosed any direct evidence to produce a definitive answer.

SOLDIERS WHO PERISHED DURING THE BATTLE

Mathias Cooper
Thomas P. Fowle
John Hale
George Lamb

SOLDIERS WHO PERISHED LATER FROM WOUNDS SUSTAINED IN BATTLE*

Lemuel Blakey
Benjamin Brigham
James Cooper
Giles Giddings
Dr. William Motley
Ashley Stephens
Leroy Wilkinson

* Olwyn Trask is generally counted among the fatalities of the Battle of San Jacinto. It is known, however, that he died from wounds sustained during the skirmish the day prior to the Battle. For purposes of this work, we have opted to include only those who died during the Battle proper or from wounds sustained during same.

ADAMS, Thompson— Company F, First Regiment Texas Volunteers. He was born in Mississippi in 1810. He came to Texas in 1834. He was a single man.[15] He received one league of land in McMullen and McGloin's Colony in July, 1835. This was in Live Oak County east of the Nueces, and one league above the mouth of the Rio Frio.[16] After the Battle of San Jacinto he received a bounty of 320 acres of land.[17] He also received one third league of land from the Liberty County board.[18]

ALEXANDER, J. B.— Company D, First Regiment Texas Volunteers. He arrived in Texas in January, 1832.[19] He received a donation of land after the Battle of San Jacinto,[20] and also received one-third league of land issued by the Fayette County board.[21]

ALEXCIN, H. Malena— First Company, Second Regiment Texas Volunteers. No information.

ALLEN, Captain John M.— Acting Major of the Regulars. Born in Kentucky, he entered the United States Navy at an early age, but left it to engage in the Greek Revolution.[22] He came to Texas in 1830.[23] In November, 1831, he received one fourth league of land in Austin's Third Colony in Brazoria County, on Chocolate Bayou.[24] He commanded a Cavalry Company at the Battle of San Jacinto and received a donation of 640 acres after the Battle.[25] He also received one twelfth league of land issued by the Austin County Board.[26] He was the first Major of Galveston, an office to which he was repeatedly re-elected. After annexation he was appointed United States Marshall, an office he held when he died, February 12, 1847.[27]

ALLISON, J. C.— First Sergeant, Company A, First Regiment Texas Volunteers. Came to Texas before May 30, 1835.[28] He joined the army in December, 1835.[29] He received one third league of land issued by the Brazoria County board after the Battle of San Jacinto.[30]

ALLISON, Moses— Company F, Second Regiment Texas Volunteers. He came to Texas before May 2, 1835. He fought as a private in the Battle of San Jacinto. After the Battle he received one-third league of land issued by the Brazoria County board.[31] He also received a donation of 640 acres.[32]

ALLSBURY, Horatio A.— Cavalry Corps. He came to Texas before August 3, 1824. He received one and one half labors of land in August, 1824.[33] He received the following grants after the Battle of San Jacinto: A donation of 640 acres, a bounty of 640 acres, and one league and one labor from the Brazoria County board.[34]

ALLSBURY, Young P.— Cavalry Corps. He arrived in Texas in 1834. He fought as a private in the Cavalry Corps in the Battle of San Jacinto. He received one-third league of land issued by the Brazoria board after the Revolution.[35] He also received a bounty of 320 acres and a donation of 640 acres.[36]

ANDERSON, John D.— Company D, First Regiment, of Texas Volunteers. He came to Texas before March 2, 1836.[37] He received a donation of 640 acres after the Battle of San Jacinto.[38]

ANDERSON, Washington— Company C, First Regiment Texas Volunteers. He was born in Virginia[39] and came to Texas in February 1835. He received two-thirds of a league and one labor of land from the Bastrop County board after the Revolution.[40] He also received a donation of 640 acres and a bounty of 320 acres of land. [41]

ANDREWS, Micah— First Lieutenant, Company C, First Regiment Texas Volunteers. He was born in Alabama in 1809 and came to Texas in February, 1835. He was without a family and was a farmer.[42] He joined the army at the beginning of the Revolution, and was made a lieutenant March 1, 1836.[43] He received a donation of 640 acres after the Battle of San Jacinto.[44]

ANGEL, John— Company B, Volunteers. He arrived in Texas in 1835. He received one-third of a league of land from the Harris County board after the Battle of San Jacinto.[45] He also received a donation of 640 acres.[46]

ARMOR, William L.— Company I, Volunteers. Came to Texas before March 2, 1836.[47] He received a bounty of 320 acres.[48]

ARMSTRONG, Irwin— Company A, First Regiment Texas Volunteers. Arrived in Texas before March 2, 1836.[49] He joined the army in January, 1836.[50] He received one-third league from the Brazoria County board after the Texas Revolution.[51] He also received a bounty of 960 acres.[52]

ARNOLD, Hayden— Captain, Company I, Second Regiment Texas Volunteers. Born in Tennessee in 1805, he came to Texas January 14, 1836.[53] He joined the army January 14, 1836.[54] After the Revolution he represented Nacogdoches in the first Congress. [55] He received one league and one labor of land from the Nacogdoches County board after the Battle of San Jacinto.[56] He also received a donation of 640 acres.[57]

ARRIOLA, Simon— 9th Company, Second Regiment Texas Volunteers. He was a native of Bexar. After the Revolution he received one league and one labor from the Bexar County board.[58] In addition, he received a bounty of 320 acres and a donation of 640 acres.[59]

ATKINSON, M. B.— 4th Company, Second Regiment Texas Volunteers. He came to Texas before the Declaration of Independence, on March 2, 1836. He received one-third league of land from the Brazoria County board after the Revolution.[60] He also received a donation of 640 acres.[61]

AVERY, Willis— Company C, First Regiment Texas Volunteers. He was born in Missouri in 1807. He arrived in Texas in December, 1831, with his wife and two sons and settled in Austin's Colony.[62] He received one league of land on Willbarger's Creek in Williamson County on November 13, 1832.[63] He got one league of land from the Bastrop County board after the Revolution.[64] He also received a donation of 640 acres and a bounty of 320 acres.[65]

BAILEY, Alex— 4th Company, Second Regiment Texas volunteers. He was born in Ohio in 1797 and came to Texas in 1827. He was a farmer.[66] He received one-fourth league of land in Austin's second colony in March, 1831. This was in Washington County, near the head waters of the best fork of Mill Creek.[67] He received one-twelfth league from the Austin County board after the Revolution.[68] He also received a donation of 640 acres of land.[69]

BAILEY, Howard— First Company, Second Regiment Texas volunteers. He came to Texas during or before 1832 and settled in Nacogdoches. He was a bachelor.[70] He joined the army March 6, 1836.[71] He received one-third league of land from the Nacogdoches County board after the Revolution.[72]

BAIN, Nicholas M.— Company C, First Regiment Texas volunteers. He arrived in Texas in 1836. He received one-third league of land from the Bastrop County board after the Revolution.[73] He also received a donation of 640 acres of land.[74]

BAKER (or BARKER), Elias— Company K, First Regiment Texas Volunteers. He was a member of Captain Calder's Company but was transferred to Captain Kuykendall's Company at Harrisburg. On November 26, 1838 he was issued 640 and 320 acres of land. He was issued one-third of a league of land by the Montgomery County Board of Land Commissioners.

BAKER, D. D. D.— Company D, First Regiment Texas volunteers. Born in Massachusetts in 1806. He came to Texas in February, 1831 and was unmarried.[75] May 2, 1831, he received one-fourth league of land in Austin's second Colony. This was in Wharton County, near the head of Bay Prairie.[76] After the Revolution he got three-fourths of a league and one labor of land from the Matagorda County board.[77] He also received a bounty and a donation of land.[78] In 1836, he was a representative from Matagorda County.[79]

BAKER, Joseph— He came from Maine to Texas, December 7, 1831. He was unmarried, and a teacher by profession.[80] He got one-fourth league of land in Austin's Fifth Colony, October 5, 1835.[81] He received one-twelfth league of land issued by Bexar County after the Revolution.[82] He also received a bounty of 320 acres of land for his services in the army.[83] He was County Judge of Bexar County in 1836.[84]

BAKER, Mosely— Company D, First Regiment Texas Volunteers. A native of Virginia,[85] he came from Alabama to Texas March 2, 1835. He had a wife and one child. He was a lawyer.[86] He received one league in Zavala's Colony, on the east shore of Galveston Bay October 9, 1835.[87] He was one of the first to raise a company for the Campaign in 1836, and one of those ordered arrested by Ugantechea at San Felipe in July, 1835. It was Baker's company that offered effectual resistance to Santa Anna, and prevented him from crossing the Brazos at San Felipe. Baker's company behaved with distinguished gallantry at San Jacinto.[88] He received a donation of land after the Battle.[89] He represented Galveston in the Congress of the Republic in 1838-39. He died of yellow fever in Houston, November 4, 1848.[90]

BALCH, H. B.— Wyly's Company.[91] He received a bounty of 320 acres and a donation of 640 acres of land for his services during the Revolution.[92]

BALCH, John— Wyly's Company. Came to Texas in April, 1835. He fought as a private in the Battle of San Jacinto. He received one-third league of land from the Matagorda County board after the Revolution.[93] He also received a donation and a bounty of land.[94]

BANKS, R.— Company I, Volunteers. He received a bounty of 320 acres of land for his services in the army.[95]

BARDWELL, S. B.— Artillery Corps. He received a bounty of land after the Battle of San Jacinto.[96]

BARKER, George— 5th Company, Second Regiment Texas Volunteers. He came to Texas in 1834. He received one-third league of land issued by the Harris County board.[97]

BARKLEY, J. A.— 4th Company, Second Regiment Texas volunteers. He arrived in Texas before the Declaration of Independence.[98] He received a bounty and a donation of land after the Battle of San Jacinto.[99]

BARR, Robert— 4th Company, Second Regiment Texas volunteers. He was born in Ohio in 1802 and came to Texas before December 5, 1832. He was unmarried.[100] He was postmaster general under President Houston in 1836-38.[101]

BARSTOW, Joshua (or Joseph)— Company B, Volunteers. Born in Boston, Massachusetts, he arrived in Texas January 28, 1836, having been recruited by Captain Turner. It is believed that Barstow was a guard at the home of Dr. J. A. E. Phelps in Brazoria County where Santa Anna was a prisoner. Barstow died at Dr. Phelps' home. His heirs did not apply for any of the land due him for his headright or for his services in the army.

BATEMAN, William— 8th Company, Second Regiment Texas Volunteers. He came to Texas in 1833. He received one-third league of land issued by the San Augustine County board.[102] He also received a donation of land after the Battle of San Jacinto.[103]

BARTON, Jefferson— Company C, First Regiment Texas Volunteers. He came to Texas in 1830 and received one-third of a league of land issued by the Bastrop County board after the Revolution.[104]

BARTON, Wayne— Company C, First Regiment Texas Volunteers. He came to Texas in 1829 and received one-third league of land issued by the Bastrop County board.[105]

BAXTER, Montgomery— Artillery Corps. He received a bounty and donation of land after the Battle of San Jacinto.[106]

BAYLOR, Dr. J. W.— 4th Company, Second Regiment Texas Volunteers. He received a bounty of 640 acres of land after the Battle of San Jacinto.[107]

BEACHAM, John— Company D, First Regiment Texas Volunteers. He came to Texas in May, 1835 and received one league of land issued by the Jasper County board.[108] He received a bounty of 320 acres for his services in the army.[109]

BEAR, Isaac— Company A, Regulars. He came to Texas on or before February 3, 1836. He received one-third league of land from the Harris County board.[110] He also received a bounty of 1280 acres after the Battle of San Jacinto .[111]

BEARD, A. J.— He came to Texas in April, 1831 and received one-third league of land from the Fort Bend County board after the Revolution.[112] He also received a donation of 640 acres.[113]

BEASON, Leander— Company F, First Regiment Texas Volunteers. He came to Texas before the Declaration of Independence. He received one-third league of land from the Colorado County board after the Revolution.[114] He also received a donation.[115]

BEAUFORD, Thomas Y.— He was in Texas before June 3, 1835. He was a bachelor.[116] September 4, 1835, he received ten leagues from Grant, Durst and Williams; this was in Goliad County, south of the San Antonio River.[117] He received one league and one labor of land from the Nacogdoches County board after the Revolution.[118] He also received a donation and a bounty of land.[119]

BEBEE, Elnathan— Company A, Regulars. He was in Texas before the Declaration of Independence. He received one-third league of land in Liberty County after the Battle of San Jacinto.[120]

BELCHER, J.— Company K, First Regiment Texas Volunteers. (Given only in the Muster Rolls in the Land Office.) He came to Texas before the Declaration of Independence. He received three-fourths of a league and one labor of land from the Washington County board after the Battle of San Jacinto.[121] He also received a donation of land.[122]

BELDEN, John— Company B, Volunteers. A native of New York, he was a member of the Volunteer Grays from New Orleans, which arrived in Texas, October 25, 1835.[123] He received one-third of a league of land issued by the Harris County board after the Revolution.[124]

BELKNAP, Thomas— 3rd Company, Second Regiment Texas Volunteers. He came to Texas in 1831. He received one-third of a league of land in 1836.[125] He also received a bounty of land after the Revolution.[126]

BELL, James— Company B, First Regiment Texas Volunteers. He came from Georgia in 1824. He was a bachelor.[127] On December 8, 1835, he received ten leagues of land from Williams, Johnson and Peebles. This was on the waters of the Sabine and Trinity.[128] He received a bounty of 320 acres of land after the Battle of San Jacinto.[129]

BELL, Peter Hansborough— Cavalry. He was a native of Virginia. He landed on Velasco in March, 1836, and walked up to Groce's, where the army was then encamped. He fought as a private at San Jacinto.[130] He received a bounty of 1280 acres of land.[131] In 1845 he was a Captain of the Rangers. He was a Colonel of Volunteers during the Mexican War and Governor of Texas from 1850 to 1853. After, he represented the Western district two terms in the United States Congress. At the expiration of his second term, he married and settled in North Carolina.[132]

BELL, Thomas H.— Company D, First Regiment Texas Volunteers. He came to Texas in January, 1834. He was a farmer.[133] On February 23, 1836, he received one league in Austin's Fifth Colony. This was in Burleson County, and on the waters of the Yegua.[134] He received one league of land from the Austin County board after the Revolution.[135] He also received a donation of land.[136]

BENCROFT, Benjamin— 5th Company, Second Regiment Texas Volunteers. No information.

BENNETT, D. W.— 8th Company, Second Regiment Texas Volunteers. He came before the Declaration of Independence. He received one-third league of land after the Battle of San Jacinto.[137]

BENNETT, Joseph H.— Lieutenant Colonel in the Second Regiment Texas Volunteers. He arrived in Texas in 1834.[138] He was a Lieutenant-Colonel in the Battle of San Jacinto. He received one-third league of land issued by the Montgomery County board after the Revolution.[139] He also received a donation and bounty of land.[140] In 1842 he raised a battalion for the expedition under Somervell; but when they reached the Rio Grande River, by permission of the Commander, Bennett and about two hundred of his men returned to their homes. He died in Navarro County in 1849.[141]

BENNET, William— 5th Company, Second Regiment Texas Volunteers. He received a bounty of 1280 acres of land after the Battle of San Jacinto.[142]

BENSON, Ellis— (From Turner's Company) Artillery. He was in Texas before March 2, 1836. He received two-thirds of a league of land from the Fort Bend County board[143] and also received a donation of 640 acres of land.[144]

BENTON, Alfred— Artillery Corps. Was born in Kentucky in 1816. He came to Texas January 9, 1836. He brought some slaves with him.[145] He received a bounty and donation of land after the Battle of San Jacinto.[146]

BENTON, Jesse— Company A, Regulars. He came to Texas before the Declaration of Independence. He received one-third league of land from Sabine County board after the Revolution.[147] He also received a bounty of 640 acres of land.[148]

BERNARD, Jason H.— Company B, Volunteers. He arrived in Texas before the Declaration of Independence. He received one-third league of land from the Fort Bend County board after the Revolution.[149] He received in addition a bounty of 960 acres and a donation of 640 acres.[150]

BERNBECK, William— Company D, First Regiment Texas Volunteers. Born in Germany, December 12, 1802, he emigrated to the United States in 1831 and settled in Pittsburgh. He enlisted in Captain William H. Smith's Company April 2, 1836 and was in Captain Baker's Company at San Jacinto. He died in 1837 and was single. He did not apply for headright or bounties.

BERRY, Andrew Jackson— Company C, First Regiment Texas Volunteers. He came to Texas in 1828. He received one-third league of land from the Washington County board after the Battle of San Jacinto.[151] He received an additional 640 acres as a donation.[152]

BERRYHILL, William H.— Company A, First Regiment Texas Volunteers. Came to Texas before the Declaration of Independence. He received one league and one labor of land from the Shelby County board.[153] In addition to this he received a donation of 640 acres and a bounty of 960 acres of land.[154]

BIGLEY, John— Company F., First Regiment Texas Volunteers. No information.

BILLINGSLY, Jesse— Captain Company C, First Regiment Texas Volunteers. He came to Texas in 1835. He received one-third league of land from the Bastrop County board after the Battle of San Jacinto.[155] In addition he received a bounty and donation of land.[156] In 1838 and the following years he commanded a ranging company upon the frontier. After annexation, Captain Billingsly represented Bastrop County in the Legislature.[157]

BINGHAM, M. A.— Company K, First Sergeant, First Regiment Texas Volunteers. He arrived in Texas before March 2, 1836. He received one-third league of land from the Harris County board.[158] He received a bounty of 320 acres in addition.[159]

BISSETT, Robert B.— Company B, Volunteers. He came to Texas in January, 1836. He received one-third league of land from the Harris County board after the Battle of San Jacinto.[160] He received a donation of 640 acres of land in addition.[161]

BLACKWELL, Thomas— Cavalry Corps. He came to Texas before the Declaration of Independence. After the Revolution he received one league and one labor, issued by the Brazoria County board.[162] In addition to this he received a donation of 640 acres.[163] He was recording clerk in the House of Representatives in the first session of Congress.[164]

BLAKEY, Lemuel S.— Company C, First Regiment, of Texas Volunteers. He came to Texas in January, 1832.[165] He was killed in the Battle of San Jacinto.[166] His heirs received one-third of a league of land from the Bastrop County board after the Revolution.[167]

BLEDSOE, George L.— 4th Company, Second Regiment Texas

Volunteers. He was in Texas before November 13, 1834. He had a wife and one son.[168] He received one-third league of land, issued by the Brazoria County board after the Revolution.[169] He received an additional 640 acres as a bounty.[170]

BLUE, Uriah— Company A, First Regiment Texas Volunteers. He came to Texas before the Declaration of Independence. He received one-third league of land from the Washington County board after the Revolution,[171] and an additional 1280 acres for services in the army.[172]

BOLLINGER, E.— 3rd Company, Second Regiment Texas Volunteers. He came to Texas in January, 1835. The Washington County board granted him one-third of a league of land after the Revolution,[173] and he received a donation of land in addition.[174]

BOLLINGER, P.— 3rd Company, Second Regiment Texas Volunteers. He came to Texas January 3, 1835. He received one-third league of land from the Washington County board after the Revolution,[175] and he received, in addition to this, a donation of 640 acres of land.[176]

BOND, H.— Company I, Volunteers. Here before February 16, 1836. He was a married man.[177] He received a bounty of 320 acres after the Revolution.[178]

BOOKER, Shields— Assistant surgeon in the Second Regiment, Medical Staff. Came to Texas before March 2, 1836. The Brazoria County board issued him one-third of a league of land after the Revolution,[179] and he received a bounty and donation of land in addition.[180]

BORDEN, John P.— First Lieutenant Company D, First Regiment Texas Volunteers. He was a native of New York and came to Texas in 1828.[181] November 20, 1832, he received one-fourth league in Austin's Second Colony. This was in Wharton County on the banks of the Colorado River.[182] He received three-fourths of a league of land from the Harris County board after the Revolution,[183] and a donation and bounty of land in addition to this.[184] He was first Commissioner of the General Land Office.[185]

BORDEN, Paschal P.— Company D, First Regiment Texas Volunteers. A native of Indiana, he came to Texas on December 17, 1829.[186] March 4, 1831, he received one-fourth league of land in Austin's Second Colony, in Washington County, on the west fork of Hill Creek.[187] After the Revolution he received three-fourths of a league of land from the Fort Bend County board.[188] In addition to this he received a bounty of 320 acres of land.[189]

BOSTIC, S. R.— Company D, First Regiment Texas Volunteers. Came to Texas in 1831. He got one-third league and one labor of land from the Austin County board after the Revolution.[190] In addition he received a bounty and a donation of land.[191]

BOTTSFORD, Seymour— Company A, First Regiment Texas Volunteers. A native of Mississippi,[192] he came to Texas before the Declaration of Independence[193] and joined the army April 4, 1836.[194] After the Revolution he received one-third of a league from the Fort Bend County board, [195] a donation of 640 acres and a bounty of 960.[196]

BOWEN, R.— Company H, First Regiment Texas Volunteers. He was a native of Tennessee. He came to Texas April 5, 1835. He left his family in Tennessee.[197] He received a bounty of 320 acres of land after the Battle of San Jacinto.[198]

BOX, James E.— First Company, Second Regiment Texas Volunteers. He came to Texas in January, 1834,[199] and on May 9, 1835 he received one-fourth league in Burnet's Colony in Grimes County.[200] He joined the army March 6, 1836.[201] He received 369 acres of land from the Houston County board,[202] and a donation and bounty of land after the Revolution.[203]

BOX, John— First Company, Second Regiment Texas Volunteers. He came to Texas before November 28, 1834, and brought his family.[204] July 30, 1835 he received one league in Vehlein's Colony in Houston County, east of the Trinity River.[205] He joined the army March 6, 1836.[206] After the Revolution he received a donation of land,[207] and one labor from the Houston County board.[208]

BOX, Nelson— First Company, Second Regiment Texas Volunteers. A bachelor who came to Texas before June 1, 1835.[209] On March 15, 1835, he received one-fourth league in Vehlein's Colony in Houston County.[210] He joined the army March 6, 1836.[211] He received a donation of land[212] as well as two-thirds of a league and one labor from the Houston County board after the Revolution.[213]

BOX, Thomas G.— First Company, Second Regiment Texas Volunteers. He came to Texas in 1834. He joined the army March 6, 1836.[214] He received two-thirds of a league and one labor of land from the Houston County board after the Battle of San Jacinto.[215] He received a donation and a bounty of land in addition to this.[216]

BOYD, J. C.— 4th Company, Second Regiment Texas Volunteers. He arrived in Texas before 1835. He received one-third of a league from Bexar County after the Revolution.[217] Beside this, he received a bounty of 640 acres of land.[218] He was a Representative in Congress from Sabine County in 1836.[219]

BOYLE, William— Fifth Company, Second Regiment Texas Volunteers. A native of Pennsylvania, he was a member of the New Orleans Grays, which arrived at Velasco October 25, 1835.[220] He received a certificate for land.

BRADLEY, Isaac— Company B, Volunteers. He arrived in Texas before the Declaration of Independence. He received one-third league of land from the Austin County board after the Revolution.[221] He received a donation and bounty in addition to this.[222]

BRADLEY, James— Fourth Company, Second Regiment Texas Volunteers. He came from Tennessee to Texas in 1831.[223] February 15, 1836, he received one-fourth league in Austin's Fifth Colony in Bastrop County, on Buckner's Creek.[224] He received the following land after the Revolution: one league and one labor issued by Sabine County,[225] and a donation of 640 acres.[226]

BRAKE, Michael J.—Third Company, Second Regiment Texas Volunteers. He arrived before the Declaration of Independence.

He received one-third league from the Jefferson County board after the Revolution.[227]

BRANCH, Edward T.—Third Company, Second Regiment Texas Volunteers. He arrived in Texas in 1835 and received one-third league of land from the Liberty County board after the Revolution.[228] He also received a donation of 640 acres.[229] He represented Liberty County in the first Congress, October, 1836, and was re-elected in 1837.[230]

BREEDING, Fidelie—Company F, First Regiment Texas Volunteers. He arrived in Texas in February, 1833. He received one-third league of land from the Fayette County board after the Revolution.[231] In addition to this he received a donation of 640 acres.[232]

BREEDLOVE, A. W.—Fourth Company, Second Regiment Texas Volunteers. He was born in New Orleans in 1799 and arrived in Texas in February, 1831. He was a farmer.[233] April 20, 1831 he received one league of land in Austin's Second Colony.[234] After the Revolution he received the following additional land: one labor from the Brazoria County board,[235] and a bounty and a donation.[236]

BRENNAN, William—Fourth Company, Second Regiment Texas Volunteers. He received a bounty of 1280 acres after the Revolution.[237]

BREWER, H. M.—First Company, Second Regiment Texas Volunteers. He came to Texas before September 20, 1834. He was a single man.[238] After the Revolution he received one league and one labor of land from the Nacogdoches County board.[239] He received a donation of 640 acres in addition.[240]

BREWSTER, Henry P.—Company A, Regulars. He came to Texas before March 2, 1836. He received one labor from the Nacogdoches County board after the Revolution.[241] He received a donation and bounty of land in addition to this.[242]

BRIGHAM, Benjamin R.—Company K, First Regiment Texas Volunteers. He came to Texas before May 2, 1835.[243] He was killed in the Battle of San Jacinto.[244] His heirs received a donation and a bounty of land,[245] as well as one-third of a league from the Brazoria County board.[246]

BRIGHAM, Moses W.—Company I, Volunteers. He came to Texas in January, 1836.[247] He received a bounty of 320 acres of land after the Battle of San Jacinto.[248]

BRISCOE, Andrew—Captain Company A, Regulars. He came to Texas in 1833. He received one league and one labor of land from the Harris County board after the Battle of San Jacinto.[249] He received a donation and a bounty in addition.[250] He was Chief Justice of Harris County for a number of years. After this, he engaged in the mercantile business in New Orleans where he died.[251]

BROOKFIELD, Francis L.—Company F, First Regiment Texas Volunteers. He came to Texas in November, 1835. He received one-third of a league from the Harris County board after the Battle of San Jacinto.[252] He received a donation in addition to this.[253]

BROOKS, Thomas D.—First Company, Second Regiment Texas Volunteers. He came to Texas before May 2, 1835. He received a league and labor of land from the Liberty County board after the Revolution.[254] He received a donation of land in addition to this.[255]

BROWN, David—Eighth Company, Second Regiment Texas Volunteers. He came from Virginia to Texas in June, 1833. He was a married man.[256] January 15, 1835, he received one league of land in Zavala's Colony in Jefferson County.[257] After the Revolution, he received one labor from the San Augustine County board,[258] and a bounty of 320 acres.[259]

BROWN, George—Company B, Volunteers. He came to Texas in 1827. May 22, 1827 he received one league in Austin's First Colony in Fort Bend County.[260] He received one league and one labor issued by the Brazoria County board after the Revolution.[261]

BROWN, Bernett E.—Company C, First Regiment Texas Volunteers. No information.

BROWN, Oliver S.—Third Sergeant, Company A, First Regiment Texas Volunteers. He came to Texas in December, 1835 and joined the army.[262]

BROWNING, George W.—Company B, Volunteers. He arrived in Texas April 6, 1835. He received one league and one labor of land from the Houston County board, after the Revolution.[263] He received in addition a donation and bounty.[264]

BRUFF, Christopher C.—Company A, Regulars. He came to Texas before March 2, 1836. He received one-third of a league of land from Harris County after the Battle of San Jacinto,[265] and a donation and bounty of land in addition.[266]

BRYAN, Luke—Third Company, Second Regiment Texas Volunteers. Arrived in Texas in 1834. He received one league and one labor from the Liberty County board.[267] He received a donation and bounty of land also.[268]

BRYAN, Moses A.—Third Sergeant, Company D, First Regiment Texas Volunteers. He arrived in Texas in 1831. In 1835 he was private secretary, first to General Austin, and afterward to General Burleson.[269] October 20, 1835, he received one-fourth league in Austin's Fifth Colony in Washington County.[270] After the Battle of San Jacinto he received 360 acres from the Brazoria County board,[271] and a donation and bounty after the Revolution.[272]

BRYANT, Benjamin—Captain Seventh Company, Second Regiment Texas Volunteers. He came to Texas in 1834 and received one league and one labor of land from the Robertson County board after the Revolution,[273] and a donation of 640 acres of land.[274]

BUFFINGTON, Andrew—Eighth Company, Second Regiment Texas Volunteers. He was born in Tennessee in 1810 and arrived in Texas January 10, 1836.[275] After the Revolution he received a donation[276] and one league and one labor of land from the Washington County board.[277]

BUNTON, John W.—Company C, First Regiment Texas Volunteers. He came to Texas in 1833.[278] April 8, 1835, he received one-fourth league in Milam's Colony in Walker County.[279] He also received a donation and a bounty of land.[280] He represented Bastrop County in the First Legislature, 1836.[281] He then represented Austin County in 1838.[282]

BURCH, James—Eighth Company, Second Regiment Texas Volunteers. He came to Texas before the Declaration of Independence. He received one-third league of land from the San Augustine board after the Battle of San Jacinto.[283] He also received a donation and bounty of land in addition.[284]

BURCH, R. O.—Eighth Company, Second Regiment Texas Volunteers. No information.

BURLESON, Aaron—Company C, First Regiment Texas Volunteers. He came to Texas in 1831 and he received one-third league of land from the Bastrop County board after the Revolution.[285] He also received a donation of 640 acres of land.[286]

BURLESON, Edward—Colonel in the First Regiment Texas Volunteers. He was born in North Carolina in 1798. When a mere lad, he joined his father's company in the War of 1812. He kept the muster roll of the Company, and thus received his first lesson in military life. He removed with his family to Missouri, where he was elected Lieutenant Colonel of the militia in 1817. He next removed to the western district of Tennessee, where he was elected Colonel of a regiment of militia. In 1831, he removed to Texas and settled in Bastrop County.[287] April 4, 1831, he received one league in Austin's Second Colony in Polk County.[288] Burleson was elected Colonel of the First Regiment in 1836.[289] After the Battle of San Jacinto, he received a bounty of land[290] and one labor from the Bastrop County board.[291] In 1837 he was elected Brigadier General of the militia, and in 1838 was appointed Colonel in the regular army. In 1841 he was elected Vice-President of the Republic and in 1843 he was a candidate for the Presidency, but was defeated by Anson Jones. Burleson was in Mexico during the Mexican War, on the staff of General Henderson. He was elected to the Senate soon after his return to Texas. At the close of the first term he was re-elected again to the Senate, but his health was declining and he died in Austin, December 26, 1851.[292]

BURT, Samuel P.—Company I, Volunteers. He arrived before the Declaration of Independence. He received one-third league of land from the Fort Bend County board.[293]

BURTON, Isaac W.—Cavalry Corps. He came to Texas in 1831.[294] On September 18, 1835, he received one league in Burnet's Colony, in Polk County.[295] He also received one labor from the Nacogdoches County board.[296] He was elected to the Senate from Nacogdoches in September, 1837, and was a member of the fourth Senate which met in 1839.[297]

BUST, Luke W.—Musician, Company A, First Regiment Texas Volunteers. Born in Illinois, he arrived in Texas in January, 1836 and died June 1836, still in the service of the army. His heirs received one-third league of land with Dana Sherman as Administrator.[298] They also received a donation and bounty.[299]

BUTTS, A. J.—Company K, First Regiment Texas Volunteers. He came to Texas before the Declaration of Independence. He received one-third league of land from the Red River County board.[300] He also received a donation and a bounty of land.[301]

BYRD, James—Company F, First Regiment Texas Volunteers. He came to Texas in 1832. He received one league and one labor of land from the Washington County board after the Revolution.[302] He also received a bounty and a donation of land.[303]

CADDELL, Andrew—Eighth Company, Second Regiment Texas Volunteers. He was a native of North Carolina. He arrived in Texas before April 3, 1834. He had a wife and eight children.[304] April 14, 1835, he received one league in Zavala's Colony, in San Augustine County.[305] He received a donation of land after the Battle.[306]

CAGE, B. F.—Fourth Company, Second Regiment Texas Volunteers. He arrived in Texas in 1832 and was a bachelor.[307] He received a donation of 640 acres after the Revolution.

CALDER, Robert J.—Captain, First Regiment Texas Volunteers. He was a native of Kentucky. He came to Texas April 16, 1833.[308] He was unmarried. He received a bounty of 640 acres of land after the Revolution.[309] Calder was sheriff of Brazoria County during the Republic. After annexation he moved to Fort Worth and became Chief Justice of the County.[310]

CALDWELL, Pinkney—Quarter-Master, Staff of Command. He was born in Kentucky in 1795. He came to Texas December 1, 1830.[311] September 8, 1835, he received three leagues in Grant, Durst and William's Colony.[312]

CALLAHAN, Joseph—Company B, Volunteers. He came from Louisiana to Texas in 1833. He had a wife and six children.[313] He received a donation of 640 acres after the Battle of San Jacinto.[314]

CAMPBELL, Michael—Artillery Corps. He joined Captain Teal's Company on August 31, 1836, which was attached to Captain Moreland's Artillery Company at San Jacinto. He was issued 1280 acres of land November 23, 1838. He did not apply for a headright or the land due him for having served.

CAMPBELL, Joseph—Fifth Company, Second Regiment Texas Volunteers. He received a donation of land after the Battle of San Jacinto.[315]

CAMPBELL, Taylor—Third Sergeant, Company C, First Regiment Texas Volunteers. No information.

CANNON, W. J.—Company H, First Regiment of Texas Volunteers. He arrived in Texas before the Declaration of Independence. He received one league and one labor from the Brazoria County board after the Revolution.[316] He also received a donation of land.[317]

CARNAL, Patric—Third Company, Second Regiment Texas Volunteers. He arrived in Texas before April 17, 1835.[318] He was a single man. September 17, 1835, he received one league of land in Robertson's Colony, in Burleson County west of the Brazos on the

San Antonio Road.[319] After the Battle he received a bounty of 320 acres of land.[320]

CARPENTER, John—Cavalry. He arrived in Texas in 1835. He received one league and one labor from the San Augustine board after the Revolution.[321]

CARPENTER, J. W.—First Company, Second Regiment Texas Volunteers. A native of Tennessee,[322] he came to Texas in January, 1836.[323] He joined the army January 14, 1836.[324] After the Battle, he received a donation of 640 acres.[325] He also received one league and one labor of land from the Nacogdoches County board.[326]

CARPER, WILLIAM M.—Surgeon, Staff of Command. He came to Texas before December, 1835. He received two-thirds of a league of land from the Harris County board.[327] He also received a donation of 640 acres.[328]

CARTER, James—Company D, First Regiment Texas Volunteers. He arrived in Texas in January, 1833. He received one-third league of land from the Nacogdoches County board after the Battle of San Jacinto.[329]

CARTER, R. W.—Second Lieutenant, Company I, Volunteers. He was born in Alabama in 1793. He arrived in Texas in April, 1831. A father five sons and three daughters, he was a farmer.[330]

CARTWRIGHT, Matthew—Second Company, Second Regiment Texas Volunteers. He arrived in Texas before May, 1835.[331] He joined the army on May 12, 1836.[332] On July 17, 1835, he received one-fourth league of land in Zavala's Colony in San Augustine County.[333] After the Battle he received two-thirds of a league and one labor of land from Montgomery County,[334] and he also received a donation of 640 acres of land.[335]

CARTWRIGHT, William—Second Company, Second Regiment Texas Volunteers. He arrived in Texas before the Declaration of Independence[336] and he joined the army on April 12, 1836.[337] He received a bounty of 320 acres after the Battle.[338]

CARUTHERS, Allen—Company H, First Regiment Texas Volunteers. He came to Texas May, 1835. He received one-third league from the Washington County board.[339] He also received a donation after the Battle of San Jacinto.[340]

CASSADY, J. W.—Company A, Regulars. A native New Yorker, he came to Texas before November, 1834.[341] He joined the army March 19, 1836[342] and on November 24, 1834, he received one-fourth league of land from Power and Hewetson in Bee County.[343] After the Battle he received a bounty of 1280 acres.[344]

CASTLEMAN, Jacob—Company H., First Regiment Texas Volunteers. He came to Texas before the Declaration of Independence. After the Battle of San Jacinto he received one-third league of land from Washington County.[345] He also received a donation of 640 acres of land.[346]

CEPTON, Cyrus—Fifth Company, Second Regiment Texas Volunteers. No information.

CHADDUCK, Richard—First Sergeant, Sixth Company, Second Regiment Texas Volunteers. He came to Texas in December, 1835. He received one-third of a league of land from the Montgomery County board after the Revolution.[347] He also received a donation of 640 acres.[348]

CHAFFIN, J. A.—Eighth Company, Second Regiment Texas Volunteers. He arrived in Texas April 29, 1834 as an unmarried man.[349] He received a donation of land after the Battle of San Jacinto.[350]

CHAMBERLAIN, Willard—Company H, First Regiment Texas Volunteers. He was born in Ohio. As a member of the New Orleans Grays he arrived in Texas October 25, 1835.[351] He received a donation of land after the Revolution.[352]

CHEVERS, John—Fifth Company, Second Regiment of Texas Volunteers. He arrived in Texas before May 25, 1835. He was married.[353] He received a donation of land after the Battle of San Jacinto.[354]

CHEVIS, John F.—Fifth Company, Second Regiment Texas Volunteers. He arrived in Texas before May 25, 1835. He was a bachelor.[355] He received a donation of land after the Battle of San Jacinto.[356]

CHILES, Lewis L.—Fourth Company, Second Regiment Texas Volunteers. Born in Virginia in 1811, he came to Texas via Tennessee in 1833. He was a member of Captain Patton's Columbia Company at San Jacinto. On April 20, 1835 he received one-fourth of a league of land in Robertson's Colony situated in Milam County. He was issued, on January 26, 1838, one-twelfth of a league of land by the Washington County Board. On June 16, 1838, he received a donation of 640 acres of land for having participated in the Battle. On March 28, 1838, he received a bounty of 640 acres.

CHOAT, David—Third Company, Second Regiment Texas Volunteers. Born in Louisiana, he arrived in Texas before January, 1835. He was single.[357] August 12, 1835, he received one league of land in Vehlein's colony.[358] He received a donation of land after the Battle of San Jacinto.[359]

CHRISTIE, John—Private Company B., Volunteers. He received a donation of land after the Revolution.[360]

CLAPP, Elisha—Private, Cavalry Corps. He arrived in Texas before April 8, 1834. He had a wife and four children.[361] On January 14, 1835, he received one league of land in Burney's Colony.[362] He received a donation of land after the Revolution.[363]

CLARK, Charles A.—Corporal, Volunteers, Company B. A native of New York. He was in the Mier Expedition in 1842.[364]

CLARKE, James—Seventh Company, Second Regiment Texas Volunteers. He was born in Mississippi, 1788, and came to Texas in 1828. He had six sons and one daughter.[365] He received a donation of land after the Battle of San Jacinto.[366]

CLARKE, John—Fifth Company, Second Regiment Texas Volunteers. He was born in Mississippi, and arrived in Texas on June 16,

1828 with his family.[367] After the Revolution he received a bounty of 320 acres.[368]

CLARKSON, James—Private in Company B, Volunteers. He arrived in Texas in December, 1835. He received one-third league of land issued by Harris County board.[369] He also received a donation of land after the Battle of San Jacinto.[370]

CLAYTON, Joseph—Private from Turner's Company, Artillery Corps. Born in Marshall County, Tennessee December 30, 1817, he came to Texas early in 1835. He was a member of Captain Turner's Company at San Jacinto. After the Revolution he returned to Tennessee. He was issued 1280 acres of land November 17, 1837 but the Land Office records do not show that he applied for the land due him for having participated in the Battle.

CLELLENS, John J.—Eighth Company, Second Regiment Texas Volunteers. He received a bounty of land of 960 acres after the Revolution.[371]

CLEMENTS, Lewis C.—Company H, First Regiment Texas Volunteers. He arrived in Texas in 1835 and received one-third league of land issued by Washington County.[372] After the Battle of San Jacinto he received a donation of land.[373]

CLEVELAND, H. M.—Sergeant-Major, First Regiment Texas Volunteers. He was born in Alabama in 1800, and arrived in Texas on October 3, 1835. He had a family of four sons.[374] After the Revolution he received a bounty of land.[375]

CLOPPER, Andrew M.—Private, Cavalry. He arrived in Texas before the Declaration of Independence. He received one-third league of land in Harris County before the Revolution.[376] He also received a bounty of 640 acres of land after the Revolution.[377]

COBBLE, Adam—Seventh Company, Second Regiment Texas Volunteers. He came to Texas in December, 1835. After the Battle of San Jacinto he received one league and one labor issued by Fort Bend County.[378] After the Battle of San Jacinto he also received a donation of land.[379]

COFFMAN, Elkin—First Company, Second Regiment Texas Volunteers. He arrived in Texas before the Declaration of Independence.[380] After the Battle he received a donation of 640 acres.[381] He also received one-third league of land from the Brazoria County board after the Battle of San Jacinto.[382]

COKER, John—Cavalry Corps. He was born in Alabama and in 1834 he came to Texas. He was a single man and a blacksmith.[383]

COLE, David—Third Company, Second Regiment Texas Volunteers. He arrived in Texas in 1830[384] and after the Battle of San Jacinto he received one-third league of land from the Jefferson County board.[385] He also received a donation of land.[386]

COLE, James—Third Company, Second Regiment Texas Volunteers. He was a native of Louisiana and had a wife and ten children.[387] He arrived in Texas on October 29, 1834.[388]

COLEMAN, Robert M.—Volunteer Aid. He was born in Kentucky in 1798. In May, 1831, he came to Texas. He was a married man and had two sons and two daughters. He was a farmer.[389] On February 1, 1835, he received twenty-four labors in Robertson's Colony in Jefferson County, south of Sabine Bay.[390] After the Battle he received a donation of 640 acres of land.[391] He was a member of the Convention in 1836. Late in 1836, President Burnet placed him in charge of a ranging force. Afterward he was removed by General Houston. In 1837 he was drowned while bathing at the mouth of the Brazos. The County of Coleman was named in his honor.[392]

COLES, B. L.—Company B, Volunteers. He came to Texas in February, 1836. After the Battle of San Jacinto he received one-third league of land from the Harris County board.[393] He also received a donation after the Battle.[394]

COLLARD, Job S.—Second Company, Second Regiment Texas Volunteers. He was living in Texas before May, 1835.[395] He joined the army March 12, 1836.[396] On May 28, 1835, he received one-fourth league of land in Vehlein's Colony in Montgomery County, on the east bank of the San Jacinto River.[397] After the Battle he received a donation.[398]

COLLICOATE, J. B.—Fifth Company, Second Regiment Texas Volunteers. After the Battle of San Jacinto he received a donation.[399]

COLLINGSWORTH, James—Aid-de Camp. He was a native of Tennessee. He filled the office of United States District Attorney before coming to Texas in 1835.[400] He received a donation of land after the Battle of San Jacinto.[401] He was made Secretary of State May 5, 1836.[402] He was a member of the Convention in 1836, and after the adjournment was sent as a Commissioner to the United States. In 1838, he was appointed Chief Justice of the Republic, and about the same time became a candidate for President. During the canvass he drowned himself in Galveston Bay.[403]

COLLINS, Willis—Artillery Corps. He arrived in Texas December 5, 1829. He was a married man and he had one son and one daughter. He owned eight slaves.[404] After the Battle he received a donation of 640 acres.[405]

COLTON, William—Volunteers. He came to Texas before the Declaration of Independence. After the Battle he received one-third league of land from the Harris County board.[406]

CONLEY, Preston—Company C, First Regiment Texas Volunteers. He came to Texas before March 2, 1836. After the Battle he received one-third league of land from Bastrop County.[407] He also received a donation of land.[408]

CONN, Joseph—Company B, Volunteers. He arrived in Texas before the Declaration of Independence. He received one-third league of land from Harris County after the Battle.[409] He also received a donation.[410]

CONNELL, Sampson—Company C, First Regiment Texas Volunteers. He came to Texas in 1834 and after the Battle of San Jacinto he received one league and one labor from the Bastrop County board.[411] He also received a donation.[412]

CONNOR, S.—Company K, First Regiment Texas Volunteers. No information.

CONNOR, Thomas—Company K, First Regiment Texas Volunteers. He arrived in Texas before 1834. On September 28, 1834, he received one league of land in Power and Hewitson's Colony in Refugio County.[413] After the Battle he received a donation of 640 acres.[414]

COOK, Francis J.—Company K, First Regiment Texas Volunteers. After the Battle of San Jacinto he received a donation of 640 acres of land.[415]

COOK, James R.—First Lieutenant Cavalry Corps. He arrived in Texas in December, 1835. After the Battle he received one-third league and one labor from the Washington County board.[416]

COOKE, William G.—Assistant Inspector-General. He came to Texas from Virginia November 8, 1835.[417] He received a donation of land after the Battle of San Jacinto.[418] In 1837, he commenced the drug business in Houston. In 1839 he was Quartermaster-General; in 1840, one of the Commissioners sent by President Lamar with the Santa Fe Expedition. In 1844 he represented Bexar County in Congress. After annexation he was Adjutant General during Henderson's Administration. He died at Seguin in 1847.[419]

COOKE, Thomas—Company K, First Regiment Texas Volunteers. He came to Texas before March 2, 1836. After the Battle of San Jacinto he received one-third league of land from Montgomery County.[420] He also received a donation of 640 acres.[421]

COOPER, (First name unknown)— Company I, Volunteers. No information.

COOPER, Mathias—Fifth Company, Second Regiment Texas Volunteers. He was killed in the Battle of San Jacinto.[422] His heirs received a bounty of 640 acres of land.[423]

CORNICAN, Michael—Third Company, Second Regiment Texas Volunteers. He arrived in Texas before the Declaration of Independence. He received one-third league from the Harris County board after the Revolution.[424] He also received a donation of land.[425]

COHRY, Thomas F.—Fourth Company, Second Regiment Texas Volunteers. He received a donation of 640 acres of land after the Revolution.[426]

CORZINE, Hershel—Eighth Company, Second Regiment Texas Volunteers. He arrived in Texas before the Declaration of Independence. He received one-third of a league issued by the San Augustine County board after the Revolution.[427] He also received a donation of land.[428]

COX, Lewis—Second Company, Second Regiment Texas Volunteers. He came to Texas before November 16, 1834.[429] July 4, 1835, he received one league in Austin's Third Colony in Walker County.[430]

COX, Thomas—Fifth Company, Second Regiment Texas Volunteers. He was born in Alabama in 1785. He arrived in Texas in March, 1822.[431] October 30, 1830, he received one league in Austin's Third Colony in Walker County.[432] He received a donation of land after the Battle of San Jacinto.[433] He went on the Mier Expedition in 1842.[434]

CRADDOCK, John R.—Company H, First Regiment Texas Volunteers. He came to Texas in June, 1833. He received one-third of a league of land from the Milam County board after the Revolution,[435] and a donation of 640 acres in addition.[436]

CRAFT, James A.—Company C, First Regiment Texas Volunteers. He came to Texas in January, 1835. He received one-third league from the Bastrop County board after the Revolution,[437] and a donation of 640 acres.[438]

CRAFT, Russell B.—First Sergeant Company C, First Regiment Texas Volunteers. He came to Texas in 1835 and received one-third league issued from Bastrop County after the Revolution.[439] He received an additional 640 acres as a donation.[440]

CRAIG, Harry C.—Corporal Company A, Regulars. He came to Texas in January, 1836 and enlisted in Captain Teal's Company of Regulars January 19, and was on that day promoted to Corporal. He was made Second Sergeant May 1, 1836, and was discharged on February 28, 1837, in order that he might be appointed Captain. On May 10, 1837, President Houston requested the Senate to confirm the appointment Craig as Captain, stating that he held such rank at that time. He was issued, on February 2, 1838, one-third of a league of land by the Harrisburg County board. Captain Craig was also issued 1280 acres of land, though he did not apply for the donation land due him for having participated in the Battle of San Jacinto.

CRAIN, Joel B.—Eighth Company, Second Regiment Texas Volunteers. He was a native of Tennessee. He came to Texas before May 24, 1835. He was a single man.[441] He received a donation of 640 acres of land after the Battle of San Jacinto.[442]

CRAVENS, Robert M.—Company C, First Regiment Texas Volunteers. He came to Texas before May, 1831.[443] May 30, 1831 he received one league in Austin's Second Colony in Fayette County.[444] After the Revolution he received one labor of land from the Brazoria board.[445] He received an additional 640 acres as a donation.[446]

CRAWFORD, Robert—Company K, First Regiment Texas Volunteers. He was born in South Carolina in 1816.[447] He arrived in Texas January, 1836. He received one-third league of land issued by the Washington County board after the Revolution.[448] He also received a bounty of land in addition.[449]

CRISWELL, William—Company C, First Regiment Texas Volunteers. He came to Texas in February, 1830. He received one-third league of land from the Fayette County board after the Revolution,[450] and an additional 640 acres as a donation.[451]

CRITTENDEN, Robert—Company I, Volunteers. He came from Kentucky.[452] He arrived in Texas in December, 1835. He received one-third league from the Harris County board after the Revolution.[453] He went on the Mier Expedition in 1842.[454]

CROSBY, G.—Sixth Company, Second Regiment Texas Volunteers. He was born in Tennessee in 1813.[455] He arrived in Texas in January, 1836 [456], and joined the army on January 14, 1836.[457] He received one-third league of land from the Montgomery County board after the Revolution.[458]

CRUNK, Nicholas—Company H, First Regiment Texas Volunteers. He came to Texas in December, 1833. He received three-fourths of a league of land from the Washington County board after the Revolution.[459]

CRUZ, Antonio—Ninth Company, Second Regiment Texas Volunteers. He came to Texas before the Declaration of Independence. He received one league and one labor from the Bexar County board after the Battle of San Jacinto.[460] He also received a donation of 640 acres of land.[461]

CUMBERLAND, George—Private from Teal's Company, Artillery Corps. He arrived in Texas in December, 1835.[462] He joined the army January 1, 1836.[463] He received one league from the Montgomery County board after the Revolution.[464]

CUMBO, James—Company A, First Regiment of Texas Volunteers. He received 320 acres of land as a bounty after the Revolution.[465]

CUNNINGHAM, L. C.—Company C, First Regiment Texas Volunteers. He was born in Tennessee in 1810 and arrived in Texas

August 1, 1835.[466] He received one-fourth league of land in Milam's Colony, in Bastrop County.[467] He received a donation of land after the Revolution.[468]

CURTIS, Hinton—Fourth Company, Second Regiment Texas Volunteers. He arrived in Texas in August, 1824 and received one league of land in Austin's First Colony in Matagorda County.[469] He received one labor of land from the Matagorda County board after the Revolution,[470] and a donation of land in addition.[471]

CURTIS, James—Company C, First Regiment Texas Volunteers. Born in Tennessee in 1809, he arrived in Texas February 2, 1835,[472] and received one-fourth league of land in Milam's Colony, in Hays County.[473] He received 960 acres as a bounty after the Revolution.[474]

CURVIER, Antonio—Ninth Company, Second Regiment Texas Volunteers. He was a native of Bexar. He received one-third league of land from the Bexar County board after the Revolution,[475] and in addition to this he received a donation of 640 acres of land.[476]

CURVIER, Matias—Ninth Company, Second Regiment Texas Volunteers. He was a native of Bexar. He received one league and one labor of land from Bexar County after the Revolution,[477] and a donation of 640 acres in addition.[478]

DALE, Elijah V.—Company A, First Regiment Texas Volunteers. He came to Texas before March 2, 1836[479] and joined the army April 10, 1836.[480] After the Battle he received one-third league from the Brazoria County board.[481] He also received a donation.[482]

DALLAS, W. R.—Company H, First Regiment Texas Volunteers. He arrived in Texas May 1, 1835. After the Battle he received one-third league from the Washington County board.[483] He also received a donation of land.[484]

DALRYMPLE, John—Company B, Regulars. He came to Texas, January, 1836. After the Battle he was issued one-third league of land by the Harris County board.[485] He also received a donation.[486]

DARE, George—Company A, Regulars. He arrived in Texas in February, 1829. After the Battle he received one league and one labor of land from Washington County.[487] He also received a bounty of 1280 acres.[488]

DARLING, John Socrates—Sixth Company, Second Regiment Texas Volunteers. He came to Texas March 26, 1835. He was married and a farmer.[489] February 13, 1836, he received one league of land in Austin's Fifth Colony between Turtle Bayou and Galveston Bay.[490] After the Battle he received a bounty of 320 acres.[491] He also received a donation and bounty.[492]

DAVIDSON, Dr. John—Surgeon, First Regiment, Volunteers. He arrived at Nacogdoches, Texas, on January 8, 1836. At San Jacinto he was Surgeon of the First Regiment Texas Volunteers, and on September 27, 1850, he was issued 640 acres of land for having participated in the Battle. He was issued, on February 10, 1853, one-third of a league of land and was issued 320 acres in addition.

DAVIS, Abner C.—Company A, First Regiment Texas Volunteers. He arrived in Texas before the Declaration of Independence.[493] He joined the army February 27, 1836.[494] After the Battle he received one-third league from Washington County.[495] He also received a bounty of 640 acres.[496]

DAVIS, G. W.—Company D, First Regiment Texas Volunteers. He arrived in Texas before September 1831, and received one league of land in DeWitt's Colony.[497] He joined the army January 14, 1836.[498] After the Battle of San Jacinto, he received one-third of a league of land from the Washington County board.[499]

DAVIS, J. K.—Fourth Company, Second Regiment Texas Volunteers. He arrived in Texas December, 1834. After the Battle of San Jacinto he received three-fourths of a league of land and one labor from the Fort Bend County board.[500]

DAVIS, J. P.—Cavalry Corps. He arrived in Texas in 1834. After the Battle he received one-third league of land from the Bastrop County board.[501] He also received a donation of 640 acres of land.[502]

DAVIS, Moses—Company H, First Regiment Texas Volunteers. He arrived in Texas before the Declaration of Independence. After the Battle he received two-thirds of a league and one labor of land from the Washington County board.[503]

DAVIS, Samuel—Company D, First Regiment Texas Volunteers. He was born in Missouri in 1793. He was married and was a farmer by occupation. He had a family of five sons and one daughter. He owned four slaves.[504] On April 7, 1835, he received one-fourth league of land in Zavala's Colony in Orange County, on the east bank of the Neches River.[505] After the Battle he received one-third league from the Jackson County board.[506] He also received a donation of land in addition.[507]

DAVIS, W. H.—Eighth Company, Second Regiment Texas Volunteers. He came to Texas March, 1835. After the Battle of San Jacinto he received one-third league of land from the San Augustine County board.[509] He also received a donation.[510]

DAVIS, Travis—Fifth Company, Second Regiment Texas Volunteers. He received a donation of 640 acres of land after the Battle of San Jacinto.[508]

DAVY, Thomas—Company C, First Regiment Texas Volunteers. He came to Texas February 8, 1833. After the Battle he received one-fourth league of land from the Montgomery County board.[511] He also received a donation.[512]

DAWSON, Nicholas—Second Lieutenant, Company B, Volunteers. He came to Texas before independence was declared.[513] After the Battle he received one-third league of land from Bexar County.[514] He also received a donation.[515] In 1842 he raised a company to reinforce the Texans in the West. The day after the Battle between Woll and Caldwell, Dawson was killed while he was trying to lead his men to the Texas Camp.[516]

DAY, F. H. R.—Company I, Volunteers. He received a donation of 640 acres after the Battle of San Jacinto.[517]

DEADRICK, DAVIS—Company K, First Regiment Texas Volunteers. After the Battle of San Jacinto he received a bounty of 960 acres.[518]

DEADRICK, GEORGE—Cavalry Corps. In 1860, his heirs received a donation of 640 acres[519] and a bounty of the same.

DEADRICK, Joel—Sixth Company, Second Regiment Texas Volunteers. No information.

DEBONNER, Dr. Tobias—Wyly's Company. He came to Texas after the Declaration of Independence.[520] After the Battle of San Jacinto he received one-third league of land in Austin County.[521] In addition he received a donation.[522]

DENHAM, M. H.—Company K, First Regiment Texas Volunteers. Born in Tennessee, he is shown as being Lieutenant in command of Captain Splane's Company at muster April 14, 1836. Captain Splane and most of the men of his company remained at the camp opposite Harrisburg to guard the baggage but Denham participated in the Battle of San Jacinto. He was issued 320 acres of land in his name on April 23, 1838. He did not apply for his headright or donation lands and was in Memphis, Tennessee, June 1, 1836, where he wrote a detailed account of the Battle in a letter to a friend.

DENMAN, Colden—Fourth Company, Second Regiment Texas Volunteers. A native of Mississippi, he came to Texas in 1834. He was a single man.[523] After the Battle he received one-third league from the Brazoria County board.[524]

DENNIS, Thomas M.—Company C, First Regiment Texas Volunteers. He arrived in Texas in 1835. After the Battle he received one-third league from the Matagorda County board.[525] He received a donation in addition.[526]

DEVERE, Cornelius—Third Company, Second Regiment Texas Volunteers. He arrived in Texas before the Declaration of Independence. After the Battle of San Jacinto he received one-third league of land from the Liberty County board.[527] He also received a bounty of 320 acres.[528]

DEWITT, James—Second Company, Second Regiment Texas Volunteers. He joined the army April 1, 1836.[529] He received a donation of land after the Battle of San Jacinto.[530]

DEXTER, Peter B.—Company D, First Regiment Texas Volunteers. He came to Texas before the Declaration of Independence. He received one-third league of land from the Harris County board after the Revolution.[531]

DIBBLE, Henry—Company K, First Regiment Texas Volunteers. He was born in New York in 1805. He arrived in Texas in December, 1830. He was a bachelor and a farmer.[532] He received one-third of a league of land from the Matagorda County board after the Revolution,[533] and a bounty of land in addition.[534]

DILLARD, Abraham—Company H, First Regiment Texas Volunteers. He was born in Missouri in 1802. He arrived in Texas in 1827 and was a farmer.[535] March 20, 1831, he received one-fourth of a league of land in Austin's Second Colony.[536] After the Revolution he received three-fourths of a league and one labor of land from the Washington County board,[537] and a donation of 640 acres in addition.[538]

DIXON, James (J. H. T.)—Company B, Volunteers. He is listed on the San Jacinto rolls printed in 1836 on page 11 and 30 of the army rolls in the General Land Office as "J. H. T. Dixon." On page 221 of the rolls, however, on a revised roll of Captain Roman's Company the name appears as "James W. Dixon." Most of the men of Captain Roman's Company were recruited in New Orleans by Captain Turner and arrived January 28, 1836 on the schooner *Pennsylvania* but it is not known that he was one of them. Neither J. H. T. nor James W. Dixon received land from the Mexican Government of headright, bounty or donation land from the Republic of Texas.

DOAN, Joseph—Wyly's Company (Only given in the Muster Rolls in the Land Office). He arrived in Texas before March 2, 1836. After the Revolution he received one-third of a league of land from the Liberty County board.[539]

DOANE, Joseph—Company I, Volunteers. I believe this is the same man given in Wyly's Company, above.

DOOLITTLE, Berry—Company H, First Regiment Texas Volunteers. He came to Texas in 1834. He received one league and one labor of land from the Gonzales County board after the Revolution.[540]

DOUBT, Daniel—First Company, Second Regiment Texas Volunteers. He received a donation of 640 acres of land after the Revolution.[541]

DOUTHATT, James—Private in the Cavalry. He arrived in Texas in February, 1836. He received one league and one labor of land from the Houston County board after the Revolution.[542] He received, in addition to this, a donation of 640 acres of land.[543]

DRISCOLL, Daniel O.—Company A, Regulars. He arrived in Texas previous to the Declaration of Independence.[544] He joined the army March 5, 1836.[545] He received one-third league of land from the Victoria County board.[546] He also received a donation.[547]

DUFFEE, William—Third Company, Second Regiment Texas Volunteers. He is listed on page 16 of the San Jacinto rolls printed in 1836 as Wm. Duffee, a member of Captain Logan's Company. On page 43 of the army rolls in the General Land Office the name of Captain Logan's roll is shown as Wm. Daffie and following it is the notation: Deserted after the battle. No one by the name of Duffie or Daffie received headright, bounty or donation Certificates.

DUNBAR, William—Company B, Volunteers. He was born in Tennessee.[548] He came to Texas in February, 1836. He received one-third of a league of land from the Harris County board after the Battle of San Jacinto[549] and a donation of 640 acres in addition.[550] He went on the Mier Expedition in 1842.[551]

DUNCAN, Jacob—Cavalry Corps. He came to Texas before May 16, 1835, and received one-fourth of a league of land in Vehlein's Colony in Angelina County, two leagues west of the Angelina River.[552]

DUNCAN, John—Company D, First Regiment Texas Volunteers. He came to Texas before May, 1835. He received one league of land from the Matagorda County board,[553] and a donation of 640 acres of land after the Revolution.[554]

DUNHAM, Daniel—Company F, First Regiment Texas Volunteers. He came to Texas in January, 1835. He received one-third of a league of land from the Washington County board after the Revolution[555] and a donation of land in addition.[556]

DUNHAM, D. T.—Artillery Corps. He came to Texas before the Declaration of Independence. He received one-third league of land from the Washington County board,[557] and a donation and bounty of land after the Revolution.[558]

DUNN, Matthew—Company H, First Regiment Texas Volunteers. He arrived in Texas in 1834.[559] December 8, 1834, he received one-fourth league in Robertson's Colony in Milam County, on the west bank of the Brazos.[560] After the Battle he received one league and one labor of land from the Washington County board.[561] He also received a donation.[562]

DURST, Edward—Fourth Company, Second Regiment of Texas Volunteers. He arrived in Texas September, 1829. He received one-third league of land from the Fort Bend County board.[563] He received in addition to this a donation and a bounty of 320 acres.[564]

DURST, R. B.—He arrived in Texas May 2, 1835. He received one-third league from the Brazoria County board after the Battle of San Jacinto.[565] He also received a donation of 640 acres.[566]

DUTCHER, Alfred—Company A, Regulars. He arrived in Texas January, 1836. He received one-third league of land from the Washington County board.[567] He also received a donation.[568]

DYCKES, L. J.—Third Company, Second Regiment Texas Volunteers. He arrived in Texas in 1835. He received one-third of a league from the Jasper County board.[569]

EARLE, William—Seventh Company, Second Regiment Texas Volunteers. A native of Louisiana, he came to Texas before 1824.[570] After the Battle of San Jacinto he received one-third league of land from Sabine County.[571]

EASTLAND, William M.—Company F, First Regiment Texas Volunteers. Born in Tennessee, he came to Texas in June, 1833.[572] After the Battle, he received one league and one labor of land from the Fayette County board.[573] He also received a donation and a bounty of 320 acres.[574] He was a captain in the Mier Expedition.[575]

EDENBURG, Cristobal—Second Company, Second Regiment Texas Volunteers. He arrived in Texas on October 10, 1824 from France.[576] On March 12, 1836, he joined the army.[577] June 12, 1835, he received one league of land in Vehlein's Colony in Walker County.[578] After the Battle he received one labor of land from Montgomery County.[579] In addition he received a bounty of 320 acres and a donation.[580]

EDGAR, J. S.—Company K, First Regiment Texas Volunteers. He came to Texas before the Declaration of Independence. He received one-third league of land from Washington County,[581] as well as a donation of land.[582]

EDSON, Amos—Company B, Volunteers. He arrived in Texas before the Declaration of Independence. He received one-third league of land from Harris County.[583] In addition he received a donation and a bounty of 1280 acres.[584]

EDWARDS, E. C.—Artillery Corps. He came to Texas before the Declaration of Independence.[585] He received a donation and a bounty of 640 acres of land.[586]

EDWARDS, Isiah—First Company, Second Regiment Texas Volunteers. He came to Texas before the Declaration of Independence. He joined the army March 6, 1836.[587]

EGBERT, J. S.—Company B, Volunteers. He arrived in Texas in 1836. After the Battle of San Jacinto he received one-third league from Harris County.[588] In addition he received a donation of land.[589]

ELDRIDGE, James—Company B, Volunteers. He came to Texas before the Declaration of Independence. He received one-third league of land from Liberty County.[590] He received a donation of land in addition.[591]

ELINGER (EHLINGER), Joseph—Third Company, Second Regiment Texas Volunteers. He was born in France in 1792 and arrived in Texas in June, 1835. He was a carpenter by occupation. He was married and had a family of two sons and one daughter.[592] After the Battle he received one league and one labor of land issued by the Colorado County board.[593] In addition he received a donation and a bounty of 720 acres of land.[594]

ELLENDER, Joseph—Company F, First Regiment Texas Volunteers. He came to Texas in 1825. He received one-third league of land from Liberty County.[595] In addition he received a bounty of 320 acres.[596]

ELLIOT, Peter S.—Company A, Regulars. He was a member of Captain Teal's company of regulars at San Jacinto. He was issued 1280 acres of land, though he did not apply for headright or donation lands. He was entitled to receive a Donation Certificate for 640 acres of land for having participated in the Battle.

ELLIOTT, J. D.—Cavalry Corps. He came to Texas in 1835. He received one-third league of land from the Harris County board.[597] In addition he received a bounty of 640 acres of land.[598]

ELLIS, Willis L.—Sixth Company, Second Regiment Texas Volunteers. He was born in Tennessee in 1818. He joined the army January 14, 1836.[599] He received two-thirds of a league and one labor of land from Washington County.[600] In addition he received a donation of 640 acres and a bounty of 320 acres.[601]

ENRIQUEZ, Lucio—Ninth Company, Second Regiment Texas Volunteers. He was a native of Bexar. He received one-third league of land issued by the Bexar County board.[602] In addition he received a donation and a bounty of 320 acres.[603]

ERATH, George B.—Company C, First Regiment Texas Volunteers. He arrived in Texas in 1833.[604] July 25, 1835, he received one-fourth league of land in Burleson County.[605] After the Battle of San Jacinto he received one-twelfth league of land issued by the Milam County board.[606] He also received a donation and a bounty of 1280 acres of land.[607]

EVITTS, James—Company H, First Regiment Texas Volunteers. He came to Texas December, 1832. He received one-third league of land issued by the Washington County board.[608] In addition he received a bounty 320 acres.[609]

EWING, Dr. Alexander—Acting Chief Surgeon, Medical Staff. He arrived in Texas in 1830.[610] October 3, 1835, he received one-fourth league in Austin's Fifth Colony in Jasper County.[611] He also received a donation and a bounty of 1280 acres of land.[612]

EYLER, Jacob—Company A, First Regiment Texas Volunteers. He came to Texas in 1835.[613] He joined the army in December, 1835.[614] He received one-third league of land from the Fayette County board.[615] He also received a bounty 320 acres of land.[616]

FARLEY, Massillon—Company A, Regulars. He came to Texas before November 1, 1835. He received one league of land in Nueces

County, near the waters of Agua Dulce in November, 1835.[617] He received a donation of 640 acres after the Revolution.[618] He was elected County Judge of Milam County, December 16, 1836.[619]

FARMER, James—Company H, First Regiment Texas Volunteers. He arrived in Texas November 5, 1835.[620] He received one league and one labor of land from the Washington County board.[621] He also received a donation of 640 acres.[622]

FARRIS, Hez—Sixth Company, Second Regiment Texas Volunteers. He was born in Alabama and came to Texas in 1833 with his wife and two children.[623] He received a donation of 640 acres after the Revolution.[624]

FARRISH, Oscar—Fifth Company, Second Regiment Texas Volunteers. He came from Virginia.[625] He arrived prior to the Declaration of Independence. He received one-third league of land from the Victoria County board after the Revolution,[626] in addition to a donation of 640 acres.[627]

FARWELL, Joseph—Third Company, Second Regiment Texas Volunteers. He came to Texas in 1835. He received one-third of a league of land from Matagorda County board.[628] He also received a bounty of 320 acres.[629]

FENNELL, George—Company I, Volunteers. He received a bounty of 320 acres.[630]

FERRELL, John—Artillery Corps. He was born in Tennessee in 1819.[631] He came to Texas in January, 1835[632] and joined the army January 14, 1836.[633] He received one-sixth of a league of land from the Washington County board after the Revolution.[634]

FERRILL, William L.—Sixth Company, Second Regiment Texas Volunteers. He was born in Tennessee in 1818 and joined the Texas army January 14, 1836.[635] He received a donation of 640 acres after the Revolution.[636]

FERTILAN, William—Sixth Company, Second Regiment Texas Volunteers. He came to Texas in January, 1834. He received one league and one labor of land from the Robertson County board after the Revolution,[637] in addition to a donation.[638]

FIELDS, Henry—Company K, First Regiment Texas Volunteers. He came to Texas before the Declaration of Independence and received one-third league of land from the Harris County board after the Battle of San Jacinto.[639] He also received a bounty of 960 acres of land.[640]

FINCH, MATTHEW—First Lieutenant, Sixth Company, Second Regiment Texas Volunteers. He arrived in Texas in 1835. He received one-third of a league of land from the Montgomery County board,[641] and also received a bounty after the Revolution.[642]

FISHER, William—Second Sergeant in the Eighth Company, Second Regiment Texas Volunteers. He came to Texas in 1823.[643] On July 4, 1835, he received twenty-four labors in Robertson's Colony in Bosque County on the Brazos.[644] He received a donation of 640 acres of land after the Battle of San Jacinto.[645]

FISHER, William S—Captain Company I, Volunteers. He was born in Virginia[646] and came to Texas in 1834.[647] He represented Gonzales in the Consultation at San Felipe de Austin in 1835. He commanded a company in the Battle of San Jacinto.[648] He received one-third league of land from Harris County after the Revolution,[649] and also received a donation of land.[650] In 1837 he was Secretary of War of the Republic. In March, 1840, the time of the Council House Fight, he was in command of the armed forces at San Antonio. In 1842, Fisher was elected Captain of a company from Fort Bend County in the Somervell Campaign. When Somervell abandoned the campaign and ordered a return to Gonzalez, Fisher and about two hundred men refused. Fisher was elected commander of the group who would continue into Mexico on the Mier Expedition. He was severely wounded in the attack on Mier and died in 1845.[651]

FITCH, B. F.—Company K, First Regiment Texas Volunteers. He was born in Tennessee and came to Texas in October, 1835. He was not married.[652] He received one-third league of land from the

Milam County board after the Revolution.[653] He also received a donation of 640 acres of land.[654]

FITZHUGH, J. P. T.—Assistant Surgeon of the First Regiment, Medical Staff. He was born in Virginia in 1815,[655] and arrived in Texas in 1831.[656] He joined the army January 14, 1836.[657] He received one-third league from the Bastrop County board after the Battle of San Jacinto.[658] He also received a donation of 640 acres of land.[659]

FLICK, John—Company D, First Regiment Texas Volunteers. He arrived in Texas in March, 1835. He received one-third league of land from the Jackson County board after the Battle.[660]

FLORES, Manuel—Company A, Regulars. He was a native of Texas.[661] In December, 1835, he received one league under Radford Berry in Fannin County, on Red River above Blue Bluff.[662] He received a donation of 640 acres after the Revolution.[663] In 1839 he went to Matamoras and was appointed by General Cañalizo to visit the Indians on the Texas frontier and rouse them to hostilities. He was discovered with about twenty-five Indians on the San Gabriel River by a Company of Rangers. Flores was killed and his dispatches captured.[664]

FLORES, Martin—Company A, Regulars. He was a native of Texas. He received a bounty of 1280 acres for his service in the Texas army.[665]

FLORES, Nepomoceno—Corporal in Ninth Company, Second Regiment Texas Volunteers. He was a native of Bexar. He received one league and one labor from the Bexar County board after the Revolution.[666] He received a donation of 640 acres of land.[667]

FLOYD, Joseph—Private in the Artillery Corps. He came to Texas in January, 1836 and received one-third league of land in Washington County.[668] He received a bounty of 320 acres.[669]

FLYNN, Thomas—Company, Regulars. He received a bounty of 1120 acres after the Battle of San Jacinto.[670]

FOLEY, Steven T.—Company F, First Regiment Texas Volunteers. He was born in Alabama and came to Texas in December, 1834. He was a single man[671] and after the Battle of San Jacinto, he received a donation of 640 acres of land.[672]

FORBES, John—Commissary General. He came to Texas before April 28, 1835.[673] On May 6, 1835, he received one league in Vehlein's Colony in Houston County on White Rock Creek.[674] He received one labor of land issued by Nacogdoches County after the Revolution.[675] He also received a donation of 640 acres of land.[676]

FORD, Charles A.—Company A, Regulars. No information.

FORD, Simon P.—Company A, First Regiment Texas Volunteers. He arrived in Texas in 1830.[677] He joined the army March 10, 1836.[678] He received one league and one labor of land from the Shelby County board after the Battle of San Jacinto.[679] He also received a bounty of 320 acres of land.[680]

FORRESTER, C.—Company K, First Regiment Texas Volunteers. He received a bounty of 320 acres after the Battle.[681]

FOSTER, Anthony—Private, Cavalry Corps. He arrived in Texas in the winter of 1835. He was a member of Captain Smith's Cavalry Company at San Jacinto, and on February 13, 1860 received 640 acres of land for having participated in the Battle. His bounty and headright Certificates, if issued, are not to be found. Mr. Foster died in Panola County, Mississippi, February 8, 1874, while an honorary member of the Texas Veterans Association.

FOSTER, J. R.—Company D, First Regiment Texas Volunteers. He came to Texas in 1832.[682] November 26, 1832, he received one-fourth league of land in Austin's Third Colony, in Fort Bend County, on Pennington Lake.[683] He received one-twelfth league of land from the Austin County board after the Revolution.[684]

FOWLE, Thomas— Born in Massachusetts, he arrived in Texas May 1, 1835 and was unmarried. He was killed at San Jacinto while serving as First Sergeant in Captain Smith's Cavalry Company. His heirs were issued 640 acres of land August 4, 1841, for having participated in the Battle.

FOWLER, S. J.—Company D, First Regiment Texas Volunteers. No information.

FOWLER, T. M.—Company K, First Regiment Texas Volunteers. He came to Texas in 1835 and received one-third league of land issued by the Matagorda County board after the Battle of San Jacinto.[685] He additionally received a donation of 640 acres of land.[686]

FOYLE (or FOGLE), Andrew—Wyly's Company. No information.

FRANKLIN, Benjamin C.—Company K, First Regiment Texas Volunteers. He came to Texas before May, 1835. He received one league and one labor from the Brazoria County board after the Revolution.[687] He also received a donation of 640 acres of land.[688] In December, 1836, he was elected a Judge of the Second Brazoria District.[689] He was one of the first settlers on Galveston Island. He frequently served as District Judge, and also represented the County in the State Legislature. He was elected to the Senate from the Galveston district in 1873, but died before the Legislature met.[690]

FRAZIER, Hugh—Company D, First Regiment Texas Volunteers. He came to Texas in March, 1835. He received one-third league of land from the Victoria County board after the Revolution.[691] He also received a donation of 640 acres.[692]

FREELE, James—Second Corporal, Company D, First Regiment Texas Volunteers. He was born in Ireland and came to Texas in 1833. He was a blacksmith.[693]

FRY, B. F.—Company I, Volunteers. He came to Texas before the Declaration of Independence. He received one-third league from the Harris County board after the Revolution.[694] He also received a donation of land.[695]

FULLERTON, William—Sixth Company, Second Regiment Texas Volunteers. He came to Texas before March 2, 1836. He received one league and one labor of land from the Robertson County board after the Revolution.[696] He also received a bounty of 320 acres of land.[697]

GAFFORD, John—Company H, First Regiment Texas Volunteers. He came to Texas before September 10, 1835. He was issued twenty-two labors of land in Robertson's Colony in Milam County.[698] He also received a donation after the Revolution.[699]

GAGE, Calvin—Company C, First Regiment Texas Volunteers. He came to Texas in 1834 and received one league and one labor of land from the Bastrop County board after the Revolution.[700] He also received a bounty of 320 acres.[701]

GAINER, John M.—Private from Teal's Company. Born in New York, he came to Texas in January, 1836. He was a farmer. He enlisted at Nacogdoches January 9, 1836, in Captain Teal's Company of Regulars, but was attached to Isaac N. Moreland's Company of Artillery for the Battle of San Jacinto. He was discharged on May 18, 1837 as a member of Captain Irvine's Company. He received one-third of a league of land from the Nacogdoches County Board but did not apply for the land due him for his services in the army and sold the rights instead.

GALLAHER, Edward—Fourth Company, Second Regiment Texas Volunteers. On May 21, 1827 he received one-fourth league of land in Austin's Second colony, north of Turtle Bayou.[702] He received one league and one labor of land in Brazoria County after the Revolution.[703] He also received a bounty of 320 acres.[704]

GALLITIN, Albert—Second Company, Second Regiment Texas Volunteers. He came from Missouri in 1832 and was a single man.[705] He joined the army March 12, 1836.[706] He received one league and one labor from the Montgomery County Board.[707] He also received a donation after the Revolution.[708]

GAMMEL, William—Wyly's Company. (His name is only given in the Muster Rolls at the Land Office.) He came to Texas before the Declaration of Independence. He received one-third league of land from Harris County after the Revolution. He also received a donation and a bounty of 640 acres.[709]

GANT, W. W.—Company K, First Regiment Texas Volunteers. He arrived in Texas February 21, 1835. He received one-third league of land from the Washington County board after the Revolution.[710] He also received a donation.[711] In 1836, he represented Washington County in the first Congress and was re-elected in 1837.[712]

GARDNER, G. W.—Company D, First Regiment Texas Volunteers. He arrived in Texas after March 2, 1836. He received one-third league of land from Matagorda County after the Revolution.[713]

GARNER, John—Company A, First Regiment Texas Volunteers. He arrived in Texas in January, 1836.[714] He joined the army April 12, 1836.[715] He received one-third league from the Nacogdoches County board after the Revolution.[716]

GARWOOD, Joseph—Company C, First Regiment Texas Volunteers. He came to Texas before the Declaration of Independence. He received one-third league from Colorado County after the Revolution. He also received a bounty of 960 acres.[717]

GAY, Thomas—Company D, First Regiment Texas Volunteers. He was born in Georgia in 1800[718] and arrived in Texas in May, 1830 and was a single man.[719] June 8, 1831 he received one-fourth league of land in Austin's Second Colony.[720] He also received a donation after the Battle.[721]

GAZLEY, Dr. Thomas J.—Company C, First Regiment Texas Volunteers. He arrived in Texas January 7, 1829. He was married and had one son.[722] He received one league of land in Austin's Second colony in Fort Bend County.[723] After the Battle of San Jacinto he received one labor of land issued by the Harris County board.[724]

In 1837 he was elected to the House of Representatives from Harris County.[725]

GENTRY, Fredrick B.—Company M, First Regiment Texas Volunteers. He arrived in Texas in December, 1836 and received one-third league of land from the Washington County board after the Revolution.[726] In addition he received a donation.[727]

GIDDINGS, Giles—Company A, First Regiment Texas Volunteers. He was born in Pennsylvania in 1812. On December 10, 1835, he came to Texas[728] and on April 11, 1836 he enlisted in the army.[729] He was wounded at the Battle of San Jacinto and died soon after from the effects of the wound.[730] His heirs received a donation of land.[731]

GILBERT, John F.—Seventh Company, Second Regiment Texas Volunteers. He was a native of South Carolina and came to Texas before January, 1835. He had a wife and eight children.[732] He received one league of land in Zavala's Colony in Sabine County June 5, 1835.[733] After the Revolution he received a donation.[734]

GILL, John P.—First Lieutenant, Fifth Company, Second Regiment Texas Volunteers. He came from Alabama to Texas in May, 1832 and was a single man.[735] May 13, 1831, he received one-fourth league in Austin's Second colony.[736] After the Battle he received one-twelfth league from Brazoria County.[737] In addition he received a bounty of 320 acres.[738]

GILL, W.—Company D, Volunteers. He arrived in Texas before the Declaration of Independence. After the Revolution he received one-third league of land from the Brazoria County board[739] and received a donation.[740]

GILLASPIE, James—Captain, Sixth Company, Second Regiment Texas Volunteers. He came to Texas in 1835 and after the Battle he received one-third league from the Montgomery County board.[741] In addition he received a donation of 640 acres.[742] He was a captain in the Mexican War in 1846[743] and was killed at the Battle of Resaca de la Palma.[744]

GILLESPIE, L. J.—Company Q, Volunteers. He was born in Tennessee in 1806 and came to Texas before January 1836, and he joined the army in January, 1836.[745] After the Revolution he received a donation.[746]

GODREE, Leflore—Third Company, Second Regiment Texas Volunteers. He arrived in Texas in 1834 and received one league and one labor of land from the Jefferson County board after the Revolution.[747] He also received a donation.[748]

GOHEEN, M. R.—Sixth Company, Second Regiment Texas Volunteers. He came from Michigan and arrived in Texas December 19, 1834. He was single[749] and after the Revolution he received one-third league from the Fayette County board.[750] In addition he received a donation.[751]

GOODIN, (First name unknown)—Calvary Corps. No information.

GOODWIN, Lewis—Company C, First Regiment Texas Volunteers. He came to Texas in 1834 and received one league and one labor of land from the Bastrop County board after the Battle of San Jacinto.[752] He also received a donation.[753]

GRAHAM, John—Company H, First Regiment Texas Volunteers. He came to Texas May, 1834 and was not married.[754] November 1, 1835, he received one league of land in Zavala's Colony.[755] After the Revolution he received one-third league of land from the Washington County board.[756] In addition he received a bounty of 320 acres.[757]

GRAVES, Alexander—Company I, Volunteers. He arrived in Texas sometime between May 2, 1835 and March 2, 1836. He was a member of Captain Fisher's Company of Velasco Blues at San Jacinto. He was issued 1280 acres of land in his name January 3, 1851 for his services in the army. In the certificate, it is stated that he was discharged for disability. He was issued one-third of a league of land by the Harrisburg County Land Commissioners, but did not claim any bounties or headrights.

GRAVES, Thomas A.—Company C, First Regiment Texas Volunteers. He came to Texas in 1831[758] and on November 10, 1835, he received land in Robertson's Colony in Milam County.[759] After the Battle of San Jacinto he received three-fourths of a league and one labor of land from the Jefferson County board.[760] He also received a donation.[761]

GRAY, James—Company H, First Regiment Texas Volunteers. He was born in Alabama and came to Texas in February, 1830. He was a farmer[762] and was married with one son and one daughter. After the Battle he received one-third league of land from the Washington County board[763] and received a donation.[764]

GRAY, M. B.—Company H, First Regiment Texas Volunteers. He was a native of South Carolina and came to Texas January, 1835.[765] On October 13, 1835, he received one-fourth league of land in Austin's Fifth colony in Bastrop County.[766] After the Revolution he received a donation.[767]

GREEN, B.—Eighth Company, Second Regiment Texas Volunteers. He came to Texas in 1827[768] and on October 27, 1831 he received one league in Austin's Fifth Colony in Fayette County.[769] He received one labor of land from the Liberty County board after the Revolution.[770]

GREEN, George—Company C, First Regiment Texas Volunteers. He was a native of Maryland and came to Texas May 7, 1835. He was a farmer by occupation[771] and after the Battle he received a donation of land.[772]

GREEN, James—Company K, First Regiment Texas Volunteers. A native of Kentucky, he came to Texas in February, 1834. He was married and had three daughters. He was a farmer[773] and after the Revolution he received one league and one labor of land from the Fayette County Board.[774] He also received a donation.[775]

GREEN, Thomas—Artillery Corps. He was born in Tennessee in 1816[776] and came to Texas in December, 1835.[777] On January 14,

1836, he joined the army,[778] and after the Revolution he received one-third league of land in Fayette County.[779] In addition he received a donation.[780] He represented Bexar County in the first Congress.[781] He died in North Carolina, January 12, 1864.

GREENLAW, A.—Company D, First Regiment Texas Volunteers. He came to Texas before September 25, 1835 and was a bachelor.[782]

GREENWOOD, James—Company A, First Regiment Texas Volunteers. He came to Texas before December, 1835.[783] After the Revolution he received a bounty of 960 acres.[784]

GREER, Thomas N. B.—Artillery Corps. He came to Texas February, 1836 and after the Battle of San Jacinto he received one-third league from the Robertson County board.[785] He also received a donation.[786]

GRICE, L. B.—Fourth Company, Second Regiment Texas Volunteers. He came to Texas before the Declaration of Independence. He received a donation after the Battle of San Jacinto. In addition he received a bounty of 640 acres, and one-third league of land in Fort Bend County.[787]

GRIEVES, (First name unknown)—Company A, Regulars. No information.

GRIFFIN, William C.—Company A, First Regiment Texas Volunteers. He was born in Tennessee in 1795 and on December 20, 1835 he came to Texas. He had a family of seven children[788] and joined the army April 10, 1837.[789] After the Revolution he received a bounty of 1280 acres.[790]

GRIGSBY, Crawford—First Company, Second Regiment Texas Volunteers. He came to Texas in October 1834 and received one-third league of land from the Houston County board, after the Revolution.[791] In addition he received a donation.[792]

GROCE, Jacob—Company H, First Regiment Texas Volunteers. He came to Texas August 2, 1834 and after the Revolution he received one-third league of land from the Milam County board.[793] He also received a donation.[794]

GUSTINE, Dr. Lemuel—Cavalry Corps. He was born in Carlisle, Pennsylvania in 1816 and came to Texas in January, 1836. He was a member of Captain Karnes' Cavalry Company at San Jacinto. He did not apply for the headright bounty and donation lands due him. He died October 3, 1852.

HAGEN, Nat—Fourth Company, Second Regiment Texas Volunteers. He arrived in Texas in January, 1836 and received one-third league of land in Harris County after the Revolution.[795] He also received a bounty of 960 acres.[796]

HALDERMAN, Jesse—Company C, First Regiment Texas Volunteers. He was a native of Kentucky and came to Texas in 1831.[797] December 3, 1832, he received one-fourth league in Austin's Fourth colony in Houston County, on the Trinity River.[798] After the Battle he received a bounty of 320 acres.[799]

HALE, John C.—First Lieutenant, Seventh Company, Second Regiment Texas Volunteers. He was born in Maine and came to Texas in February, 1831.[800] He was killed in the Battle of San Jacinto.[801] His heirs received a donation of land after the Revolution.[802]

HALE, William—Company K, First Regiment Texas Volunteers. After the Revolution he received a donation.[803]

HALL, James—Fourth Company, Second Regiment Texas Volunteers. He was born in Illinois and came to Texas August 4, 1831. He was married and had a family of two sons, owned one slave and was a farmer.[804] November 6, 1835 and received one league of land in Robertson's Colony in Harrison County.[805] After the Revolution he received one labor of land from the Nacogdoches County board.[806] In addition he received a bounty of 640 acres of land.[807]

HALL, John—Company K, First Regiment Texas Volunteers. He came to Texas before April, 1831. April 23, 1831, he received one-fourth league of land in Austin's Second colony.[808] After the Battle he received a donation.[809]

HALLET, John—Company F, First Regiment Texas Volunteers. He came to Texas May 2, 1835 and received one labor of land from the Colorado County board after the Revolution.[810] He also received a donation.[811]

HALLMARK, W. E.—First Company, Second Regiment Texas Volunteers. He came to Texas in December, 1834 and received one league and one labor of land from the Houston County board after the Battle of San Jacinto.[812] He also received a donation.[813]

HALSTEAD, E. B.—Company K, First Regiment Texas Volunteers. The name "E. B. Halsey" appears on page 6 of the army rolls in the General Land Office as a member of Captain Wadsworth's Company which arrived at Velasco on December 20, 1835 and served under Fannin in 1836. Following his name is the notation: "Left." This means that he left the Company prior to February 29, 1836 when the company was mustered. The name "E. B. Halstead" is listed on page 14 of the San Jacinto rolls printed in 1836 as a member of Captain Calder's Company. It is possible that E. B. Halsey and E. B. Halstead were one and the same. No one by the name of Halstead or Halsey received headright, bounty, or donation land certificates.

HAMILTON, E. E.—First Company, Second Regiment Texas Volunteers. He came to Texas in January, 1836 and after the Revolution he received one-third league from the Nacogdoches County board.[814] In addition he received a donation.[815]

HANCOCK, George—Eighth Company, Second Regiment Texas Volunteers. He came to Texas in 1836 and after the Revolution he received one-third league of land from the Brazos County board.[816]

HANDY, Robert E.—Volunteer Aid. He arrived in Texas in July, 1834. He received one-third league from the Fort Bend County board.[817] In addition he received a donation[818] and was a merchant by occupation.[819]

HARDEMAN, Thomas M.—Company F, First Regiment Texas Volunteers. He came to Texas in 1835 and after the Battle he received one-third league from the Matagorda County board.[820] He also received a donation.[821] He was elected to the House of Representatives from Matagorda County in September, 1837.[822]

HARDIN, Franklin—First Lieutenant, Third Company, Second Regiment Texas Volunteers. He came to Texas in 1825[823] and received on May 12, 1831, one-fourth league of land on the west bank of the Brazos.[824] He received one-twelfth league from Liberty County after the Revolution.[825] In addition he received a donation.[826]

HARMON, Clark M.—Artillery Corps. He was born in Tennessee in 1817 and came to Texas in January, 1836. He joined the army January 14, 1836.[827] After the Revolution he received a donation.[828]

HARMON, John—Eighth Company, Second Regiment Texas Volunteers. He was a native of Tennessee. He came to Texas in 1830. He was a married man.[829] March 12, 1835, he received one league in Robertson's Colony in McLennan County on the Brazos.[830] After the Revolution he received a donation.[831]

HARPER, John—Company B, Volunteers. Born in 1811, he arrived at Velasco January 28, 1836, having been recruited in New Orleans by Captain Turner. He was issued, on March 2, 1838, by the Harrisburg County Board, one-third of a league of land. He was also issued 640 acres of land for having participated in the Battle of San Jacinto. He left his home on the date of his death to look after his flock of sheep and sought to drive them to shelter. The night was cold, and was supplemented later by a severe storm. Nearing a twenty foot embankment, he evidently paused, and accidently slipped down the embankment, breaking both of his legs. Searching parties found him the next morning, near to noon, frozen stiff; he'd been dead for several hours. His body was buried in the Cedar Cemetery about five miles from LaGrange.

HARPER, Benjamin J.—Second Lieutenant, Third Company, Second Regiment Texas Volunteers. He was a native of Virginia. He came to Texas before December 19, 1834. He was married.[832] March 4, 1835, he received one league in Vehlein's Colony in Polk County.[833] He received a donation in addition after the Revolution.[834]

HARPER, Peter—Fourth Company, Second Regiment Texas Volunteers. He arrived in Texas before March 2, 1836. He received one league and one labor of land from the Nacogdoches County board after the Revolution.[835] He also received a donation.[836]

HARRIS, A. J.—Company I, Volunteers. He was a native of Tennessee and arrived in Texas in February, 1831.[837]

HARRIS, James—Fourth Company, Second Regiment Texas Volunteers. He was a native of Alabama and arrived in Texas in 1830. He was married.[838] October 10, 1835, he received one league in G. W. Smyth's grant in Harrison County.[839] After the Revolution he received a donation.[840]

HARRIS, T. O.—Artillery Corps. He was born in Tennessee in 1815. January 9, 1836 he came to Texas with a family of servants.[841] After the Revolution he was granted one-third league by the Harris County board.[842]

HARRIS, William—Second Lieutenant, Cavalry Corps. He was born in Louisiana in 1799 and came to Texas in July, 1830. He was a bachelor and a carpenter by occupation.[843] December 18, 1830 he received two-thirds league in Austin's Third Colony in Brazoria County.[844] After the Revolution he received a bounty of 320 acres.[845]

HARRISON, A. L.—Second Lieutenant, Sixth Company, Second Regiment Texas Volunteers. He was a native of New York and came to Texas in 1830.[846] After the Battle of San Jacinto he received a donation.[847]

HARVEY, David—Company B, Volunteers. He was born in England. He came to Texas in September, 1834.[848] After the Battle he received a donation.[849]

HARVEY, John—First Company, Second Regiment Texas Volunteers. He was born in Alabama in 1805. He came to Texas January 7, 1836. He was married and had a son and a daughter.[850] After the Battle he received one-third league from the Nacogdoches County board.[851] He also received a donation.[852]

HARKIN, Thomas A.—Company A, First Regiment Texas Volunteers. He came to Texas before the Declaration of Independence. After the Revolution he received one-third league in Bastrop County.[853]

HASSELL, J. W.—Company K, First Regiment Texas Volunteers. He came to Texas May 2, 1835. He received one-third league from the Bastrop County board after the Revolution.[854] In addition he received a donation.[855]

HAWKINS, William—Company H, First Regiment Texas Volunteers. He was a native of Alabama and came to Texas in 1829.[856] December 20, 1832, he received one-fourth league in Austin's Second Colony. Two-thirds of his land was in Burleson County, and one-third in Washington County.[857] After the Revolution he received a donation.[858]

HAWKINS, William—Company D, First Regiment Texas Volunteers. He was a native of Missouri and came to Texas in 1832.[859] He received a donation after the Battle of San Jacinto.[860]

HAYES, C.—Company A, First Regiment Texas Volunteers. He was a native of Tennessee. He came to Texas in February, 1831. He was a farmer.[861] After the Battle he received a donation.[862]

HAYS, James—Fourth Company, Second Regiment Texas Volunteers. He came to Texas before May 2, 1835. After the Revolution he received one-third league of land from the Brazoria County board.[863] He also received a donation.[864]

HEARD, William J. E.—Captain Company F, First Regiment Texas Volunteers. He came in 1830, with a caravan of colonists, from northern Alabama to Texas.[865] November 30, 1830, he received one league of land in Austin's Third Colony in Jackson County.[866] In 1840 Captain Heard accompanied Colonel John H. Moore in a campaign against the Indians on the upper Colorado. Later he was Chief Justice of Wharton County. After the close of the Civil War, he moved to Chappell Hill, Texas, where he died in August, 1874.[867]

HECK, Charles F.—Fourth Company, Second Regiment Texas Volunteers. He arrived at Velasco, December 20, 1835 as a member of the Georgia Battalion. The name C. F. Heck appears on the roll of Captain Ticknor's Company, following which is the notation: Lost on Ward's retreat. On the rolls of Fannin's command Chas Hec is shown as unattached to any company. C. F. Hick appears as a member of Captain Ticknor's Company who was among those who escaped from Colonel Ward's Division and was not captured by the enemy. There seems to be but little doubt, if any, that Charles Hec, Charles Heck and C. F. Hicks are one and the same person. No one by the name of C. F. Heck, C. F. Hick, C. F. Hee, Charles Hee, Charles Hec, Charles F. Hicks, or Charles Frederick Hicks applied for headright, bounty or donation Certificates for the land, if any, due him.

HEISER, J. A.—Wyly's Company. (His name is given only on the Muster Rolls at the Land Office) He came to Texas before the Declaration of Independence. He received one league and one labor of land from the Brazoria County board after the Revolution.[868] In addition he received a donation.[869]

HENDERSON, Robert—Company A, Regulars. He arrived in Texas between May 2, 1835 and March 2, 1836. Major Caldwell appeared before the Victoria County board and signed an affidavit that Henderson had died at the camp at Victoria, July 20, 1836, while serving in the army. He was issued, on February 1, 1833, one-third of a league of land by the Victoria County Board of Land Commissioners delivered to Pinckney Caldwell, Administrator of his estate. The Administrator of his estate did not apply for the donation land or the bounty land due him for having participated in the Battle.

HENDERSON, F. K.—Company H, First Regiment Texas Volunteers. He was a native of Louisiana. He came to Texas in June, 1831. He was married and had two sons and three daughters.[870] After the Battle he received a donation.[871]

HENDERSON, Hugh—Cavalry Corps. He came to Texas in 1835. He received one-third league from the Montgomery County board after the Revolution.[872] In addition he received a donation.[873]

HENDERSTON (or HENDERSON), Augustus—Wyly's Company. (His name is given only in the Muster Rolls at the Land Office.) After the Battle he received a bounty of 640 acres.[874]

HENRY, Robert—Sixth Company, Second Regiment Texas Volunteers. He came to Texas November, 1834.[875] December 22, 1834, he received one league in Robertson's Colony in Live Oak County.[876] After the Battle he received a labor from the Robertson County board.[877] In addition he received a donation.[878]

HERRERA, Pedro—Ninth Company, Second Regiment Texas Volunteers. He was a native of Texas. After the Battle of San Jacinto he received one league and one labor from the Bexar County board.[879] He also received a donation.[880]

HERRING, John—Company C, First Regiment Texas Volunteers. He received a donation after the Revolution.[881]

HIKOX, Franklin—Wyly's Company. (His name is given only in the Muster Rolls at the Land Office.) After the Battle of San Jacinto he received a bounty of 320 acres.[882]

HIGHLAND, Joseph—Company F, First Regiment Texas Volunteers. He was a native of New York. He came to Texas June, 1831.[883] After the Battle he received one-third league of land from the Colorado County board.[884] In addition he received a donation.[885]

HIGHSMITH, A. M.—Company C, First Regiment Texas Volunteers. He was a native of Missouri, was married and had a family of two daughters and one son.[886] He came to Texas in 1827.[887] April 5, 1831 he received one league in Austin's Second colony in Bastrop County.[888] He received one labor from the Bastrop County board after the Revolution.[888] He also received a donation.[890]

HILL, Abraham Webb—Cavalry. After the Battle he received a donation.[891] Born in Georgia, he came to Texas in 1835. He enlisted in Captain Henry W. Karnes' company April 12 and continued in it until July 12th. He was permitted by General Thomas J. Rusk to join the rangers under Captain John G. McGehee and continued until November 12, 1836. He was issued 640 acres of land July 5, 1859, for having served in the army. He was in Captain Karnes' company at San Jacinto and on January 7, 1839 was issued 640 acres of land for having participated in the Battle. He died in Bastrop County in 1884 while a member of the Texas Veterans Association.

HILL, H.—Eighth Company, Second Regiment Texas Volunteers. Came to Texas before July 14, 1835. He received one league of land in Milam's Colony in Houston County in July, 1835.[892]

HILL, Isaac—Company D, First Regiment Texas Volunteers. He received a donation after the Revolution.[893]

HILL, James M.—Company H, First Regiment Texas Volunteers. He came to Texas in 1835. After the Battle of San Jacinto he received one-third league from the Washington County board.[894] He also received a donation.[895]

HILL, J. W.—Cavalry Corps. He came to Texas in February, 1836. After the Revolution he received one league of land from the Robertson County board.[896]

HILL, William W.—Company H, First Regiment Texas Volunteers. He came to Texas in February, 1835. After the Battle he received one-third league from the Washington County board.[897] In addition he received a donation.[898]

HOBSON, John—Company C, First Regiment Texas Volunteers. He was born in Tennessee in 1813. He came to Texas on December 10, 1835. He had a family of servants.[899]

HOCKLEY, George W.—Inspector General. He came to Texas previous to the Declaration of Independence. He received one-third of a league from the Fort Bend County board after the Revolution.[900] He also received a donation.[901] He was Secretary of War under Houston in 1843, and was sent to Mexico to negotiate a peace. He died in Houston in 1854.[902]

HOGAN, Thomas—Company B, Volunteers. He arrived at Velasco January 28, 1836, having been recruited in New Orleans by Captain Turner. He is shown as having enlisted in the army February 13, 1836, for a period of two years. At the promotion of Captain Turner to Lieutenant Colonel, the name of his company was transferred to Company A, First Regiment Regular Infantry, of which John Smith was appointed Captain, August 29, 1836. His name appears as a member of Captain Smith's Company at muster December 31, 1836. He did not apply for the land due him for having served in the army. He died at Galveston August 1, 1837, while still in the army.

HOGANS, Josiah—Company F, First Regiment Texas Volunteers. He came to Texas before December, 1834. He received one league in Robertson's Colony in Falls County west of the Brazos, December 20, 1834.[903] In addition he received a bounty of 640 acres after the Revolution.[904]

HOGG, W. C.—Company K, First Regiment Texas Volunteers. No information.

HOLDEN, Prior—Company C, First Regiment Texas Volunteers. He came to Texas in 1833. After the Battle he received one league and one labor of land from the Bastrop County board.[905] In addition he received a donation.[906]

HOLLINGSWORTH, James—Company H, First Regiment Texas Volunteers. He was born in Arkansas in 1791 and came to Texas in 1829. He was not married.[907] October 28, 1831, he received one-fourth league in Austin's Second colony in Fort Bend County.[908] After the Battle he received a donation.[909]

HOLLMAN, S.—Eighth Company, Second Regiment Texas Volunteers. He arrived in Texas before the Declaration of Independence. After the Revolution he received two-thirds league and one labor from the San Augustine County board.[910]

HOLMES, Peter W.—First Company, Second Regiment Texas Volunteers. He came to Texas in the fall of 1835. He received one-third league from the Nacogdoches County board after the Revolution.[911] In addition he received a donation.[912]

HOMAN, Harvey—Company B, Volunteers. He came to Texas before the Declaration of Independence. After the Battle of San Jacinto he received one-third league of land from the Brazoria County board. In addition he received a donation.[913]

HOOD, Robert—Wyly's Company. He is shown as a member of Captain Wyly's Company. He did not apply for headright or donation certificates. He was issued 320 acres of land in his name April 9, 1838, for his services, but the certificate had been assigned to S. Rhoades Fisher.

HOOD, Robert—Company C, First Regiment Texas Volunteers. Received a bounty of 320 acres after the Battle of San Jacinto.[914]

HOPE, Prosper—Company H, First Regiment Texas Volunteers. He was a native of Louisiana. He came to Texas in 1825.[915] December 28, 1828, he received one league of land in Austin's Second Colony in Washington County.[916] After the Revolution he received a donation.[917]

HOPE, Richard—Company H, First Regiment Texas Volunteers. He received a bounty of 320 acres and a donation after the Revolution.[918]

HOPKINS, Thomas—Fifth Company, Second Regiment Texas Volunteers. He was born in Indiana in 1802. He arrived in Texas March 7, 1823. He was a laborer.[919] After the Revolution he received one-third league from the Bastrop County board.[920]

HORTON, Alexander—Aid-de-camp. He was a native of North Carolina. He came to Texas before September 24, 1824.[921] February 24, 1835, he received one-fourth league in Zavala's Colony in Jefferson County.[922] After the Revolution he received three-fourths league and one labor from the San Augustine County board.[923] In addition he received a donation.[924]

HOTCHKISS, Rinaldo—Eighth Company, Second Regiment Texas Volunteers. He came to Texas in November, 1835. He received one-third league from the Nacogdoches County board after the Revolution.[925] He also received a donation of land.[926]

HOUSTON, Samuel—Major General commanding. He was born in Virginia in 1793. His father died when he was fourteen years old. The widow Houston and her nine children moved to Blount County, Tennessee. Young Sam spent his time alternatively at school, at farm work, and as a clerk in a store. In 1809, he ran away from home and joined a band of Cherokee Indians hunting in the neighborhood. He spent three years with them. He returned home and engaged in school teaching for awhile.

In 1813 he enlisted as a soldier in the Creek War. His gallantry won the admiration of General Jackson, who became his life-long friend. In 1817 he was appointed Indian Agent but soon resigned and began the study of law. In 1819 he was elected District Attorney of Davidson County, and at the same time Major-General of the militia. In 1823 he was elected to Congress and re-elected in 1835. At the close of his second term he was elected Governor of Tennessee. In 1829, he married young Eliza Allen. The union, still shrouded in mystery, would last but eleven weeks. The citizens of Nashville were shocked to learn that Eliza had returned to her father's house in Gallatin, and that Houston had hastily resigned as Governor and fled from the city. He took passage on a steamer and went to join his old companions among the Cherokees, then living on the Arkansas River in the Indian Territory.

On October 29, 1829 he was formally admitted to citizenship among the Cherokees. In 1832 he visited the city of Washington in the interest of the Indian tribes. He was given a commission as Confidential Indian Agent among the tribes of the Southwest, to whom he was sent to negotiate treaties. His visit to Washington revived in him a love for civilized life and he was anxious to re-enter public life. He decided to come to Texas, partly to look for a new home, and partly to fulfill his mission to the Indian tribes within her territory. He reached Texas December 10, 1832. From this period Texas became his home, and for thirty years his character reigned as her principal political and historical figure. The first service he rendered Texas was as a delegate to the Convention of 1833. In September 1835, he introduced a series of resolutions at a public meeting at Nacogdoches, to consider the possibility of convening a consultation. He was a member of the General Consultation at San Felipe, in 1835 and elected to the convention which met in March 1, 1836. The Declaration of Independence took place on the 2nd, and Houston was elected commander-in-chief of the army. Two days later he left for the army, then on the banks of the Guadalupe.

His talents, his former experience, and his splendid victory at San Jacinto, framed him as the most suitable person to fill the executive chair of the young Republic and he was elected President in September 1836. He was succeeded by Vice-President Lamar, but was re-elected President in 1841. President Houston was suc-

ceeded by Dr. Anson Jones. At the first session of the State Legislature in 1846, he was elected to the United States Senate, and was
re-elected in 1847, and again in 1851. In 1857, Houston announced
as an independent candidate for Governor. He was beaten by
Honorable H. R. Runnels. In 1859, Houston again became an independent Democratic candidate for Governor and was elected
by a large majority. Houston was not in favor of the secession of
Texas in 1861. The Convention met March 5, 1861, and passed a
bill uniting Texas to the new Confederacy. Houston refused to
take the oath to support the new government and made no serious opposition to retiring to private life. He viewed with grief the
war measures adopted by both the North and the South. His last
public appearance was before an audience in Houston, March 19,
1863. He died in Huntsville, July 26, 1863.[927]

HOWELL, Robert—Company A, First Regiment Texas Volunteers. He came to Texas in 1834.[928] He joined the army April 10,
1836.[929] After the Battle he received one-third league from the
Harris County board.[930]

HUGHES, J.—Company H, First Regiment Texas Volunteers.
(His name is only given in the Muster Rolls of the Land Office.)
He was a native of Mississippi and came to Texas in 1831.[931] June
6, 1832 he received one-fourth league in DeWitt's Colony in Gonzales and DeWitt counties, on the northeast bank of the Guadalupe.[932] After the Revolution he received a donation.[933]

HUGHES, Thomas M.—Eighth Company, Second Regiment Texas Volunteers. He was a native of Alabama. He came to Texas in
December, 1830. He was married, had two daughters and was a
farmer.[934] After the Battle of San Jacinto he received one league
and one labor of land from the Shelby County board.[935] In addition he received a donation.[936]

HUNT, John C.—Company H, First Regiment Texas Volunteers.
He was a native of Alabama and came to Texas January 1, 1835.
He was a farmer.[937] After the Battle of San Jacinto he received one-

third league of land from the Washington County board.[938] In addition he received a donation.[939]

INJAMS, Basil J.—Second Lieutenant, Fifth Company, Second Regiment Texas Volunteers. Came from Alabama to Texas February 8, 1835. He was a farmer.[940] He received one-third league from the Brazoria County board after the Revolution.[941] He received in addition a bounty of 320 acres and a donation of 640 acres of land.[942]

INGRAM, Allen—Company D, First Regiment, Texas Volunteers. He came to Texas prior to the Declaration of Independence. After the Revolution he received one-third league issued by the Washington County board.[943] He also received a bounty of 640 acres.[944]

INGRAHAM, John—Company H, First Regiment Texas Volunteers. He came with the family of Thomas Williams from Arkansas in 1821, when he was fourteen years old. He was an orphan, but he had a guardian in Arkansas. He told his guardian that he would return home in a year. He did go back to Arkansas in 1822. In 1823 he came back to Texas with William Rabb and James Gilleland. In the campaign of 1836, Ingraham was a private in Captain Hill's Company, and he did yeoman services in the Battle of San Jacinto.[945] He received one-third league of land from the Bexar County board after the Revolution.[946] He also received a donation of 640 acres, and a bounty of 320 acres of land.[947] In 1837 Ingraham married and settled on the Colorado, nine miles from LaGrange. In 1847 he was elected a Captain of Militia and was Commissioned by Governor J. Pinkney Henderson.[948]

IRVINE, James T. P.—Seventh Company, Second Regiment Texas Volunteers. He arrived in Texas in 1830 from Tennessee.[949] He received a donation of land after the Battle of San Jacinto.[950]

IRVINE, J. S.—Seventh Company, Second Regiment Texas Volunteers. He arrived in Texas in 1830. He got two-thirds of a league and one labor of land from the San Augustine County board after the Revolution.[951] He also received a bounty of 320 acres and a donation of 640 acres of land.[952]

ISBELL, James H.—Company D, First Regiment Texas Volunteers. He came to Texas previous to the Declaration of Independence. He got two-thirds of a league of land from the Liberty County board after the Battle of San Jacinto.[953] He also received a donation of 640 acres and a bounty of 640 acres.[954]

ISBELL, William—Company D, First Regiment Texas Volunteers. He came to Texas in July, 1835. After the Revolution he received one-third league of land issued by the Harris County board.[955] He also received a donation of 640 acres, and a bounty of 640 acres.[956] During the Republic, he was a member of Captain Mark B. Lewis' Ranging Company. He became blind in 1856 and died in 1877.[957]

JACK, William H.—Fourth Company, Second Regiment Texas Volunteers. He was born in Alabama and came to Texas in 1830.[958] He was a member of the Committee of Safety of Columbia in 1835, and also connected with the Army of the West the same year. He fought as a private at San Jacinto. The same year he was Secretary of State in Burnet's Cabinet.[959] He received a labor of land from the Brazoria County board after the Revolution.[960] He also received a donation of 640 acres of land.[961] He was a member of the House of Representatives in 1839 and enlisted in the army to repel the Mexican invasion in 1842.[962] He contracted yellow fever in Galveston and died in August, 1844.[963]

JACKSON, T. R.—Company D, First Regiment Texas Volunteers. He came to Texas before March, 1835. He received one league in Milam's Colony in Rusk County, on the waters of the Angelina in March, 1835.[964] He received a donation and bounty of 1280 acres after the Battle of San Jacinto.[965]

JAMES, Denward—Company I, Volunteers. He came to Texas before the Declaration of Independence and received one-third league of land from the Brazoria County board after the Battle of San Jacinto.[966] He also received a bounty of 320 acres.[967]

JAQUES, Isaac—Fifth Company, Second Regiment Texas Volunteers. He was born in New York and came to Texas in October, 1835. He was married and had two daughters. He was a farmer.[968] He received a donation of land after the Battle of San Jacinto.[969]

JENNINGS, J. D.—Company H, First Regiment Texas Volunteers. He came to Texas prior to the Declaration of Independence. After the Revolution, he received one-third league from the Colorado County board.[970] He also received a bounty of 640 acres.[971]

JETT, J. M.—Company B, Volunteers. He was born in Louisiana and came to Texas before December, 1834. He was a married man.[972] He received one league in Zavala's Colony in Orange County, on Cow Creek, February 11, 1835.[973] He received a donation of land after the Battle of San Jacinto.[974]

JETT, Steven—Company B, Volunteers. He was a native of Kentucky and went to Tennessee before coming to Texas.[975] He came to Texas in 1830. He had a wife and one child.[976] He received on February 18, 1835, one league in Zavala's Colony in Orange County, between Cow and Adams Creek.[977] He received one labor issued by Jefferson County board after the Revolution.[978] He also received a donation of 640 acres and a bounty of 1280 acres.[979]

JOHNSON, (First name unknown)—Company B, Volunteers. No information.

JOHNSON, Benjamin—Sixth Company, Second Regiment Texas Volunteers. He came to Texas in 1832. He received two-thirds of a league and one labor from the Jefferson County board after the Revolution.[980] He also received a bounty of 320 acres and a donation of 640 acres.[981]

JOHNSON, George—Cavalry Corps. He came to Texas January, 1836. He received one league and one labor issued by Nacogdoches County board.[982] He also received a donation of 640 acres.[983]

JOHNSON, George J.—Company K, First Regiment Texas Volunteers. He came to Texas before the Declaration of Independence. He received one league from the Matagorda County board and a donation of 640 acres after the Revolution.[984]

JOHNSON, James R.—Seventh Company, Second Regiment Texas Volunteers. He was a native of Virginia and came to Texas before 1834.[985] He received one-fifth league from the Brazoria County board.[986] He also received a donation of 640 acres, and a bounty of 1280 acres.[987]

JOHNSON, Thomas F.—Sixth Company, Second Regiment Texas Volunteers. He came to Texas in 1835. He received one-third league from the Red River County board.[988] He also received a donation of 640 acres and a bounty of 320 acres after the Revolution.[989]

JONES, Allen B.—Company F, First Regiment Texas Volunteers. He came to Texas July 1, 1826. He received one league and one labor from the Montgomery County board after the Revolution.[990] He also received a donation of 640 acres.[991]

JONES, Dr. Anson—Surgeon in the Second Regiment. He was born in Massachusetts in 1789. He was licensed to practice medicine in 1820. He spent two years in Virginia, but came to Texas in 1833 and settled at Brazoria. Dr. Jones enlisted as a private in Captain Calder's Company, but was soon afterward appointed surgeon in Burleson's Regiment.[992] He received one-third league and one labor from the Brazoria County board after the Revolution.[993] He also received a bounty of 1280 acres of land.[994]

In 1837 he represented Brazoria County in Congress. In 1838 he was minister to the United States, and while absent was elected to the Senate. Jones was Secretary of State during Houston's second term, and at the close of that term was elected President of the Republic. After his term of office expired, he retired to his home in Washington County and lived in private life. In 1857 he sold his home with a view of settling on the coast. On January 7, 1858, he was at the old Capitol House in Houston and said to a friend:

"Here, in this house, twenty years ago, I commenced my political career in Texas as a member of the Senate, and here I would like to close it." Not long afterward a pistol shot was heard in the room, and Dr. Jones was found in a dying condition.[995]

JONES, Edward S.—Company I, Volunteers. He received a bounty of 320 acres, and a donation of 640 acres after the Revolution.[996]

JONES, George W.—Eighth Company, Second Regiment Texas Volunteers. He was born in Tennessee and came to Texas before 1829.[997] He received one league and one labor form the San Augustine County board.[998] He also received a bounty of 320 acres and a donation of 640 acres.[999]

JORDAN, A. S.—Company B, Volunteers. He came to Texas January, 1836. He received one league and one labor from the Nacogdoches County board.[1000]

JOSLYN, (First name unknown)—Company I, Volunteers. No information.

KARNER, John—Company A, Regulars. He came to Texas in March, 1836. He received one-third league of land after the Revolution.[1001] He received 640 acres donation.[1002]

KARNES, Henry W.—Captain in the Cavalry Corps. He was a native of Tennessee. Early in life he attached himself to a company of trappers on the frontier of the Arkansas River. The company disbanded on the head of Red River. Karnes and three companions crossed the country to the Trinity River. They built a canoe and descended the stream to Robbin's Ferry. Karnes crossed over to the Brazos and found employment as an overseer on the Groce plantation. He responded to the first call for volunteers at the beginning of the Revolution and distinguished himself at the taking of the city of San Antonio. He proved one of the best cavalry scouts and spies, and commanded a company of cavalry at San Jacinto. After the Battle he went to Matamoras to effect an exchange of prisoners, and was himself thrown in prison but escaped.[1003] He received one-third league of land issued by the Bexar County board after the Revolution.[1004] He also received a donation of 640 acres.[1005] In 1837 he was an Indian Agent, and in 1838-39 he was in the Ranging service. He received a severe wound in a single combat with an Indian chief and died in San Antonio in 1840, from the effects of the wound. Captain Karnes was wholly uneducated, but he was one of a class of men to whom Texas owes a lasting debt of gratitude.[1006]

KELSO, Alfred—First Corporal, Company F, First Regiment Texas Volunteers. He was born in Tennessee and came to Texas via Alabama in February, 1835. He was a single man.[1007] He got one league and one labor of land from the Colorado County board after the Revolution.[1008] He also got a donation of 640 acres.[1009]

KENT, Joseph—Eighth Company, Second Regiment Texas Volunteers. He arrived in Texas before 1832. He received one-fourth league in DeWitt's Colony in DeWitt County in June, 1832.[1010] He received one-twelfth league from the Austin County board after the Battle of San Jacinto.[1011] He also received a donation of 640 acres.[1012]

KENYON, A. D.—Fourth Company, Second Regiment Texas Volunteers. He came to Texas before October, 1835. He received one-third league from the Colorado County board.[1013] He also received a bounty of 320 acres after the Revolution.[1014]

KIBBLE, William—Third Company, Second Regiment Texas Volunteers. He came to Texas in 1834. He received one-third league from the Liberty County board.[1015] He also received a donation of 640 acres after the Revolution.[1016]

KILLEEN, W. H.—Company B, Volunteers. He received a bounty of land after the Battle of San Jacinto.[1017]

KIMBROUGH, Captain William—Eighth Company, Second Regiment Texas Volunteers. He was born in Tennessee and came to

Texas in 1831. He had a wife and one child.[1018] He received on September 17, 1835 one league in Burnet's Colony. He received one league and one labor issued by the Shelby County board, after the Revolution.[1019] He also received a donation of 640 acres.[1020]

KINCANNON, William P.—First Company, Second Regiment Texas Volunteers. He came to Texas February 4, 1836. He received one league and one labor from the Harris County board.[1021] He also received a donation of 640 acres after the Revolution.[1022]

KING, W. P.—Cavalry Corps. He was born in Georgia and came to Texas in 1836. He had a wife and four children.[1023] He received one league and one labor from the Shelby County board.[1024] He also received a donation of 640 acres.[1025]

KLEBERG, Robert—Company D, First Regiment Texas Volunteers. He was born in Germany and came to Texas in December, 1834. He was a farmer.[1026] He received one league and one labor from the Austin County board.[1027] He received a donation of 640 acres of land after the Battle.[1028]

KORNEGAY, David S.—Company H, First Regiment Texas Volunteers. He came to Texas in 1830 and received one-third league of land in Fayette County after the Battle of San Jacinto.[1029] He also received a bounty of 320 acres.[1030]

KRAATZ, Lewis—Wyly's Company. (His name is given only in the Muster Rolls in the Land Office.) He came to Texas prior to the Declaration of Independence. He received one league and one labor from the Matagorda County board.[1031] He also received a donation of 640 acres.[1032]

KUYKENDALL, Mathew—Company D, First Regiment Texas Volunteers. He came to Texas in 1831 and received one-third of a league issued by the Austin County board.[1033] He also received a donation of 640 acres after the Revolution.[1034]

LABADIE, N. D.—Assistant Surgeon, Second Regiment, Medical Staff. He came to Texas in 1830 and after the Battle of San Jacinto he received one league and one labor from the Liberty County board.[1035] In addition he received a donation.[1036]

LABALTRIER, Charles—Company A, Regulars. He came to Texas before the Declaration of Independence. After the Revolution he received one-third league from Washington County. In addition he received a bounty.[1037]

LAMAR, Mirabeau Buonaparte—Commanding Cavalry Corps. He was born in Georgia in 1798. In early life he was private secretary to Governor Troup of that state. After the death of his wife, he traveled extensively and in 1835 followed Fannin to Texas. When heard the news of Goliad and the Alamo, he hurried to Velasco. He reached the army at Groce's Point and enlisted as a private. Lamar commanded the cavalry in the Battle of San Jacinto.[1038] He received a bounty of land after the Battle.[1039] Lamar was Secretary of War in Burnet's Cabinet. He was elected Vice-President in October, 1836 and was elected President at the end of Houston's term. At the commencement of the Mexican War, Lamar was appointed Division Inspector under General Henderson. In 1847 he was Post Commander at Laredo. On his return to Texas he was elected to the Legislature. He died on December 19, 1859.[1040]

LAMAR, S. W.—Company B, Volunteers. He received a bounty of 1280 acres.[1041]

LAMB, George A.—Second Lieutenant, Second Company, Second Regiment Texas Volunteers. He was a native of Kentucky and came to Texas in October, 1834. He had a wife and five children.[1042] He joined the army March 12, 1846.[1043] He was killed in the Battle of San Jacinto and his heirs received one league and one labor of land from the Montgomery County board.[1044] In addition they received a donation.[1045]

LAMBERT, Walter—Company K, First Regiment Texas Volunteers. Arrived in Texas in June 1834. He received one-twelfth of a league of land from the Victoria County board after the Battle of San Jacinto.[1046] In addition, he received a donation.[1047]

LANE, Walter P.—Fourth Company, Second Regiment Texas Volunteers. He was born in Ireland in 1817.[1048] March 1, 1836, he arrived in Texas. After the Revolution he received one-third league of land from the Jasper County board.[1049] In addition he received a donation.[1050] He died in 1887.[1051]

LANE, George W.—Company A, Regulars. He arrived in Texas before January, 1836. He joined the army January 20, 1836.[1052] After the Revolution he received one league and one labor of land from the Harrisburg County board.[1053] In addition he received a donation.[1054]

LAUGHRIDE, William—Company A, First Regiment Texas Volunteers. He received a donation after the Revolution.[1055]

LAURENCE, G. W.—Second Company, Second Regiment Texas Volunteers. He arrived in Texas in January, 1835.[1056] He joined the army March 12, 1836.[1057] After the Battle he received one league and one labor of land from the Montgomery County board.[1058] In addition he received a donation.[1059]

LAURENCE, Samuel—Company H, First Regiment Texas Volunteers. He was a native of Arkansas and came to Texas February 23, 1829.[1060] March 26, 1831 he received one league in Austin's Second Colony in Burleson County on the Brazos River.[1061] After the Revolution he received one labor of land from the Washington County board.[1062]

LESSITER, Francis B.—Sixth Company, Second Regiment Texas Volunteers. He received a donation and a bounty of 320 acres.[1063]

LEGG, Seneca—Private, Turner's Company. He arrived in Texas in 1833. He received one labor after the Revolution from the Montgomery County board.[1064] In addition he received a bounty of 1280 acres.[1065]

LEGRAND, E. O.—Eighth Company, Second Regiment Texas Volunteers. He was a native of North Carolina and came to Texas before December 3, 1834. He had a wife and one child.[1066] May 25, 1835, he received one league in Zavala's Colony in Jasper County.[1067] After the Revolution he received one-third league from the San Augustine County board.[1068] In addition he received a donation.[1069]

LEEPER, Sam—First Company, Second Regiment Texas Volunteers. He came to Texas in 1834 and after the Revolution he received one-third league from the Jackson County board.[1070] In addition he received a donation.[1071]

LESASSIER, Alexander—Company H, First Regiment Texas Volunteers. He arrived in Texas in 1834 and after the Battle he received one-third league of land from the Nacogdoches County board.[1072] In addition he received a bounty of 320 acres.[1073]

LESTER, James S.—Company F, First Regiment Texas Volunteers. He came to Texas in February, 1834 and after the Revolution he received one-third league from the Fayette County board.[1074] In addition he received a donation.[1075] He was a member of the first Congress from Bastrop & Gonzales Counties, as well as a member of the fourth Congress which met at Austin in November, 1839.[1076]

LEWELLAN, John—Company I, Volunteers. He received a donation and a bounty of 640 acres after the Revolution.[1077]

LEWIS, Abraham—Fourth Company, Second Regiment Texas Volunteers. After the Revolution he received one league and one labor form the Sabine County board.[1078] In addition he received a donation and a bounty of 320 acres.[1079]

LEWIS, A. S.—Second Lieutenant, Seventh Company, Second Regiment Texas Volunteers. He received a donation and a bounty of 1280 acres.[1080]

LEWIS, Edward—Company B, Volunteers. He was a member of Captain Roman's Company. Most of the men of Captain Roman's

Company arrived at Velasco, January 28, 1836 on the schooner Pennsylvania, and it is assumed, but not proven, that Mr. Lewis was among them. He served in the army from February 14 to August 14, 1836. He did not apply for the headright, bounty and donation land due him.

LEWIS, G. W.—Fourth Company, Second Regiment Texas Volunteers. He was a native of Tennessee and came to Texas in 1830 and was a farmer.[1081] After the Revolution he received one-third league from the San Augustine County board.[1082] In addition he received a donation.[1083]

LEWIS, George W.—Company I, Volunteers. He came to Texas before the Declaration of Independence. After the Revolution he received one league from Brazoria County. He also received a donation and a bounty of 320 acres.[1084]

LEWIS, John—Company F, First Regiment Texas Volunteers. He was a native of New York.[1085] He came to Texas before the Declaration of Independence.[1086] November 22, 1832, he received one-fourth league in Austin's Fourth Colony in Fayette County.[1087] After the Revolution he received one-third league in Harris County and a bounty of 480 acres.[1088]

LIGHTFOOT, William D.—Company F, First Regiment Texas Volunteers. He arrived in Texas before September 13, 1835. He received ten leagues from Williams, Johnson and Peebles.[1089] After the Revolution he received a donation and a bounty of 640 acres.[1090]

LIGHTFOOT, Wilson—Company F, First Regiment Texas Volunteers. He came to Texas in April, 1830.[1091] April 4, 1831, he received one-fourth league in Austin's Second Colony in Bastrop County.[1092] After the Revolution he received one-third league of land in Fort Bend County.[1093] He also received a donation.[1094]

LIMSKI, Frederick—Company A, Regulars. He came to Texas February, 1836. After the Revolution he received one-third league of land from the Harris County board.[1095]

LIND, John S.—Company B, Volunteers. He came to Texas in January, 1836. After the Revolution he received one-third league from the Harris County board.[1096] In addition he received a donation.[1097] He represented Victoria in 1837 in the Legislature.[1098]

LINDSAY, Benjamin—Seventh Company, Second Regiment Texas Volunteers. He was a native of Tennessee. He came to Texas September 22, 1834. He had a wife and nine children.[1099] February 21, 1835, he received one league of land in Zavala's Colony in Sabine County.[1100] After the Battle he received one-third league in San Augustine County.[1101] He also received a donation.[1102]

LODERBACK, John D.—Company A, First Regiment Texas Volunteers. He enrolled in Captain Sherman's Company in Kentucky, December 18, 1835 and arrived in Texas in January, 1836. He did not apply for land due to him for his services in the army.

LOGAN, William M.—Captain Third Company, Second Regiment Texas Volunteers. He was a native of Tennessee. He came to Texas before April 7, 1835. He was not married.[1103] November 7, 1835 he received one-fourth league in Vehlein's Colony in San Jacinto County.[1104] After the Revolution he received a donation.[1105]

LOWRIE, John—Fifth Company, Second Regiment Texas Volunteers. He arrived in Texas before the Declaration of Independence. He received one-third league in San Augustine County after the Revolution. He also received a donation and a bounty of 320 acres.[1106]

LOVE, D.—Eighth Company, Second Regiment Texas Volunteers. He arrived in Texas in 1835. After the Revolution he received one-third league in San Augustine County.[1107] He also received a donation.[1108]

LOVE, Robert—Seventh Company, Second Regiment Texas Volunteers. He came to Texas in January, 1836. After the Revolution he received one-third league of land from the Sabine County board.[1109]

LUDUS, (First name unknown)—Company B, Volunteers. No information.

LYFORD, John—Company H, First Regiment Texas Volunteers. He arrived in Texas before May, 1835. After the Revolution he received one-third league from the Washington County board.[1110] He also received a donation.[1111]

LYNCH, Nicholas—Adjutant, Staff of the Command. He was a native of Alabama and on March 22, 1828, he came to Texas. He was married and had four sons.[1112] After the Revolution he received one labor of land from the Harris County board.[1113] He also received a donation.[1114]

MADEN, Isaac—Fifth Company, Second Regiment Texas Volunteers. He was born in Indiana in 1805 and came to Texas in 1825.[1115] June 16, 1831, he received one-fourth league in Austin's Second Colony in Lavaca County.[1116] After the Revolution he received one-twelfth league from the Milam County board.[1117] He also received a donation.[1118]

MAGILL, William H.—Company C, First Regiment Texas Volunteers. He came to Texas in 1835. After the Revolution he received two-thirds league and one labor from the Bastrop County board.[1119] He also received a donation.[1120]

MALONE, Charles—Company K, First Regiment Texas Volunteers. A native of Pennsylvania, he arrived in Texas in 1834 and was a member of Captain Dimitt's Company in 1835 and of Captain Calder's Brazoria Company at San Jacinto. After the Revolution he presented no claim for the land due him as a homestead or for his services in the army.

MANCHA, Antonio—Ninth Company, Second Regiment Texas Volunteers. A married man with a family of seven children (four sons and three daughters), he was a native of Texas.[1121] December 22, 1832, he received fourteen labors in Austin's Second Colony in Austin County.[1122] After the Revolution he received one league and one labor from the Bexar County board.[1123] He also received a donation.[1124]

MANCHA, Jose Maria—Ninth Company, Second Regiment Texas Volunteers. He was a native of Texas and received one-third league and one labor from Goliad County after the Revolution.[1125]

MANUEL, Albert C.—Eighth Company, Second Regiment Texas Volunteers. A member of Captain William Kimbro's Company at San Jacinto, he did not receive headright, bounty or donation certificates. On April 29, 1836 Captain William H. Smith certified that John W. Carter lost a gun valued at $25.00 in the Battle of San Jacinto. Mr. Carter sold the certificate issued to him for that amount to Albert C. Manuel. On November 27, 1837, he received payment from the government.

MARSH, Shubael—Company A, Regulars. He came to Texas April 15, 1824.[1126] July 8, 1824, he received one league in Austin's First Colony in Brazoria County.[1127] After the Battle he received a bounty of 320 acres.[1128]

MARSHALL, John—Company D, First Regiment Texas Volunteers. He was born in Arkansas in and he came to Texas in January, 1830. He was married and had six children (five sons and one daughter) and was a farmer.[1129] After the Battle of San Jacinto he received twenty-five labors from the Milam County board.[1130] In addition he received a donation.[1131]

MASON, Charles—Company A, Regulars. He received a bounty of 320 acres of land.

MASON, George W.—Company I, Volunteers. He was born in Louisiana, 1804. He arrived in Texas in January, 1831. He was married and had three daughters.[1132] After the Revolution he received one-third league from the Harris County board.[1133] He also received a donation.[1134]

MASSIE, William—Company B, Volunteers. He came to Texas before the Declaration of Independence. After the Revolution he received one league and one labor from Red River County board. He also received a donation.[1135]

MATTY, Thomas—Company I, Volunteers. He came to Texas before the Declaration of Independence. He received one-third league from the Brazoria County board after the Revolution.[1136]

MAXWELL, P. M.—Third Company, Second Regiment Texas Volunteers. He was born in Illinois[1137] and came to Texas in 1835. He received one-third league from the Liberty County board after the Revolution.[1138] He also received a donation.[1139] He went on the Mier Expedition in 1842.[1140]

MAXWELL, Thomas—Eighth Company, Second Regiment Texas Volunteers. He was a native of Tennessee and came to Texas before March 28, 1835. He had a wife and five children.[1141] After the Revolution he received one league and one labor from the Fayette County board.[1142] He also received a bounty of 320 acres.[1143]

MAYBEE, George—Company I, Volunteers. He received a bounty of 1280 acres after the Revolution.[1144]

MAYS, Ambrose—Fifth Company, Second Regiment Texas Volunteers. He came to Texas in 1831 and after the Revolution he received one league and one labor from the Harris County board.[1145] He also received a donation.[1146]

MAYS, Thomas H.—Company C, First Regiment Texas Volunteers. He was born in Tennessee in 1803 and he came to Texas in May, 1830.[1147] April 4, 1835, he received one-fourth league in Austin's Second Colony in Bastrop County.[1148] After the Revolution he received a donation.[1149]

McALLISTER, J. D.—Company I, Volunteers. He came to Texas May 2, 1835. He was not married.[1150] After the Revolution he received one-third league from Brazoria County.[1151] He also received a donation.[1152] He went on the Santa Fe Expedition in 1842.[1153]

McALLISTER, Joseph—Company I, Volunteers. He was born in Pennsylvania in 1813 and he came to Texas in July, 1835.[1154] After the Revolution he received one-third league in Harris County.[1155] He also received a donation.[1156]

McCLELLAND, Sam—Company C, First Regiment Texas Volunteers. He was a native of Ireland[1157] and he came to Texas in 1835. After the Revolution he received one-third league from Harris County.[1158] He went on the Mier Expedition in 1842.[1159]

McCLOSKY, Robert D.—Second Lieutenant, Company A, Regulars. He was elected Second Lieutenant of the Infantry, Regular army December 17, 1835. He was Second Lieutenant of Captain Briscoe's Company at San Jacinto. He was issued 1280 acres of land on June 15, 1838, for his service in the army. The land was surveyed in the present county of Medina. He was made Captain during a recess of Congress and on May 10, 1837, President Houston requested the Senate of the Congress of the Republic to confirm his nomination. The senate did so on May 22, 1837.

McCONKLIN, Jesse L.—Company H, First Regiment Texas Volunteers. He arrived in Texas before the second of March, 1836. He received one labor of land from Washington County and a donation.[1160]

McCORLEY, Placido—Fifth Company, Second Regiment Texas Volunteers. He came to Texas before the Declaration of Independence and received one-third league from Harris County.[1161]

McCORMICK, J. M.—Fourth Company, Second Regiment Texas Volunteers. He was a native of Kentucky and came to Texas on May 24, 1831.[1162] June 22, 1832, he received one league in Austin's Second Colony in Fort Bend County.[1163] After the Revolution he received a donation.[1164]

McCOY, John—First Company, Second Regiment Texas Volunteers. He was born in Missouri and came to Texas in 1828. He was married and had three sons and one daughter.[1165] After the

Revolution he received one-third league from the Jackson County board.[1166]

McCRABB, John—Company K, First Regiment Texas Volunteers. He came to Texas in 1831, and on July 13, 1831 he received one-fourth league in DeWitt's Colony in DeWitt County on the northeast bank of the Guadalupe.[1167] After the Revolution he received three-fourths league and one labor from the Victoria County board.[1168] He also received a donation.[1169]

McCRABB, Joseph—Company D, First Regiment Texas Volunteers. He was a native of Tennessee and came to Texas in 1834.[1170] After the Revolution he received a donation.[1171]

McCULLOUGH, Benjamin—Artillery Corps. A native of Tennessee and he came to Texas at the beginning of the Revolution. He enlisted as a private but was ordered to command one of the cannons in the Battle of San Jacinto. In 1840 he represented Gonzales County in Congress, but was on the frontier as captain or in a ranging company most of the time. He was Quartermaster during the Mexican War. In 1853 he was United States Marshal of Texas. He was sent by President Buchanan in 1855 to settle a difficulty among the Mormons in Utah. He was appointed a Brigadier General in the Confederate ranks, and ordered to Arkansas. He was killed at Pea Ridge, Arkansas, March 24, 1862.[1172]

McFADDEN, David H.—Third Company, Second Regiment Texas Volunteers. He came to Texas in 1834 and after the Revolution he received one league and one labor of land from the Jefferson County board.[1173] He also received a donation.[1174]

McFARLAND, J. W. B.—Company B, Volunteers. He came to Texas in January, 1836 and after the Revolution he received one-third league in Victoria County.[1175] He also received a donation.[1176]

McGARY, D. H.—Eighth Company, Second Regiment Texas Volunteers. He came to Texas before the Declaration of Independence. He received one league and one labor in Montgomery County. He also received a donation.[1177]

McGAY, Thomas—Fourth Company, Second Regiment Texas Volunteers. No information.

McGOWN, A. J.—Eighth Company, Second Regiment Texas Volunteers. He arrived in Texas in 1835 and after the Battle of San Jacinto he received one-third league from San Augustine County.[1178] He also received a donation.[1179]

McHORSE, J. W.—First Company, Second Regiment Texas Volunteers. He received a bounty of 320 acres.[1180]

McINTIRE, Thomas H.—Captain, Fifth Company, Second Regiment Texas Volunteers. He was a native of Tennessee and came to Texas in 1829. He was not married.[1181] After the Revolution he received one league and one labor from the Jackson County board.[1182] He also received a donation.[1183]

McKAY, Daniel—Company H, First Regiment Texas Volunteers. He was born in Augusta, Maine, October 16, 1814. He came to Texas in 1834. He was a member of Captain Hill's Company. He was issued 320 acres of land for his services in the army. He was issued 640 acres of land for having participated in the Battle of San Jacinto. Daniel McKay enlisted in the Confederate Army at the outbreak of the Civil War and was stationed at Galveston. He died on his farm October 9, 1889. His grave at the Science Hill Cemetery in Bell County, seven miles east of Bartlett, Williamson County, is marked.

McKAY, Daniel—Cavalry Corps. He was born in Maine in 1812 and he came to Texas November 21, 1834. He was a bootmaker.[1184] After the Battle he received one-third league from the Washington County board.[1185] He also received a donation.[1186]

McKENZIE, Alexander—Seventh Company, Second Regiment Texas Volunteers. He arrived in Texas December, 1834 and after the Revolution he received one-third league from the Red River County board.[1187] He also received a donation.[1188]

McKENZIE, Hugh—Company F, First Regiment Texas Volunteers. He came to Texas in December, 1834. After the Revolution he received one-third league from the Red River County board.[1189] He also received a donation.[1190]

McLAUGHLIN, Robert—Company F, First Regiment Texas Volunteers. He was a member of Captain Heard's Company of Citizen Soldiers. No one by that name received headright, bounty or donation Certificates.

McLEAN, Dugald—Company C, First Regiment Texas Volunteers. He received a donation after the Revolution.[1191]

McLINN, Stephen—First Company, Second Regiment Texas Volunteers. No information.

McMANUS, R. O. W.—Third Company, Second Regiment Texas Volunteers. He came to Texas in 1835 and after the Revolution he received two-thirds league and one-labor from the Liberty County board.[1192]

McMILLAN, Edward—Sixth Company, Second Regiment Texas Volunteers. He came to Texas in June, 1834. After the Revolution he received one-third league from the Robertson County board.[1193] He also received a donation.[1194]

McNEIL, Pleasant D.—Company K, First Regiment Texas Volunteers. He received one league in Austin's First Colony August 7, 1824, east of Bernard in Brazoria County.[1195] After the Revolution he received one labor from the Brazoria County board.[1196] He also received a donation.[1197]

McNEILL, (First name unknown)—Company I, Volunteers. No information.

McNELLY, Bennett—Sergeant Major, Second Regiment Texas Volunteers. He was a native of Pennsylvania and came to Texas in 1825.[1198] After the Revolution he received one-third league from the Montgomery County board.[1199] He also received a donation.[1200]

McSTEA, Andrew—Company B, Volunteers. He came to Texas in 1836 and after the Revolution he received one-third league from the Harris County board.[1201] He also received a bounty of 1280 acres.[1202]

MENEFEE, John S.—Company D, First Regiment Texas Volunteers. He was a native of Alabama. He came to Texas in 1830. He was a farmer.[1203] After the Revolution he received one-third league from the Jackson County board.[1204] He also received a donation.[1205] He was the representative from Colorado County in 1837 and the re-elected in 1838. He was in the army to repel the Mexican invasion in 1842.[1206]

MERCER, Eli—Company F, First Regiment Texas Volunteers. He was born in Mississippi in 1785 and he came to Texas in December, 1825. He was married and had three sons and two daughters. He was a farmer.[1207] December 13, 1830, he received one league in Austin's Third Colony.[1208] In addition he received one labor after the Revolution, from the Colorado County board.[1209] He also received a donation.[1210]

MERCER, Elijah—Company F, First Regiment Texas Volunteers. He came to Texas before 1835. He was not married and a farmer.[1211] After the Revolution he received one-third league from the Colorado County board.[1212] He also received a donation.[1213]

MERCER, J. K.—First Company, Second Regiment Texas Volunteers. He came to Texas before 1833 and he was a farmer.[1214] On October 14, 1835, he received one league in Vehlein's Colony.[1215] After the Battle he received one labor from the Montgomery County board.[1216] In addition he received a donation.[1217]

MERWIN, J. W.—Private from Turner's Company, Artillery Corps. He received a donation after the Revolution.[1218]

MILES, A. H.—Company I, Volunteers. He received a donation after the Battle of San Jacinto.[1219]

MILES, Edward—Company A, First Regiment Texas Volunteers. He was born in Arkansas in 1799 and he came to Texas in January,

1830. He was a farmer and he was married. He had three daughters and one son.[1220] He joined the army on April 10, 1836.[1221] After the Revolution he received a donation,[1222] one league and one labor from the Liberty County board.[1223]

MILLARD, Henry—Lieutenant Colonel, Commander, Regulars. He came to Texas in August, 1835. After the Revolution he received one league and one labor from the Jefferson County board.[1224] He also received a donation.[1225] He died in 1842.[1226]

MILLER, Daniel—Company F, First Regiment Texas Volunteers. He came to Texas before the second of March, 1835. After the Revolution he received one-third league from the Colorado County board.[1227] He also received a donation.[1228]

MILLER, H.—Company B, Volunteers. He received a donation after the Revolution.[1229]

MILLER, James—Wyly's Company (recorded only in the Muster Rolls at the Land Office). He came to Texas in 1832 and after the Revolution he received one-third league from the Bastrop County board.[1230]

MILLER, Joseph—Company D, First Regiment Texas Volunteers. He was born in Missouri in 1792 and he was a farmer. He was married and had three sons and two daughters.[1231] April 16, 1831, he received one league in Austin's Second Colony.[1232] After the Revolution he received a bounty of 960 acres.[1233]

MILLER, William—First Lieutenant, Company B, Volunteers. He came to Texas before the Declaration of Independence. He received one league and one labor in Brazoria County after the Revolution. He also received a bounty of 1280 acres.[1234]

MILLER, William H.—Company H, First Regiment Texas Volunteers. He was a native of Kentucky and came to Texas in February, 1831.[1235] After the Revolution he received a donation.[1236]

MILLERMAN, Ira—Teal's Company, Artillery Corps. He came to Texas before the Declaration of Independence. After the Revolution he received one-third league in Montgomery County. He also received a bounty of 1280 acres.[1237]

MILLETT, Samuel—Company D, First Regiment, Texas Volunteers. He was born in Maine in 1799. He came to Texas in April, 1831.[1238] He received one league in Austin's Second Colony, November 20, 1832.[1239] After the Revolution he received a donation.[1240]

MILLS, Granville—Company K, First Regiment Texas Volunteers. He was a native of Kentucky and he came to Texas in 1831.[1241] After the Revolution he received one league and one labor from the Sabine County board.[1242]

MIMS, Benjamin—Company K, First Regiment Texas Volunteers. He arrived in Texas before the Declaration of Independence.[1243] May 22, 1835, he received one-fourth league in Milam's Colony.[1244] After the Revolution, he received a bounty of 640 acres.[1245] He also received a donation.[1246]

MINNETT, Joshua—Company B, Volunteers. He received a donation after the Revolution.[1247]

MITCHELL, Asa—Company B, Volunteers. Came to Texas before 1824. July 7, 1824, he received one-half league in Austin's First Colony.[1248] After the Revolution he received a bounty of 320 acres.[1249]

MITCHELL, James—First Regiment Texas Volunteers. He came to Texas in February, 1834.[1250] He received one-fourth league in Burnet's Colony, on April 2, 1835.[1251] After the Revolution he received one league and one labor from the Houston County board.[1252] He also received a donation.[1253]

MITCHELL, N.—Company H, First Regiment Texas Volunteers. He came to Texas before May, 1835. After the Revolution he received one-third league from the Washington County board.[1254] He also received a donation.[1255]

MITCHELL, S. B.—Company K, First Regiment Texas Volunteers. He was a member of Captain Calder's Company. No one of that name received headright, bounty or donation certificates, and no information concerning him has been found. In error, Mitchell's name was listed as Jones B. Mitchell on the bronze plaque in the San Jacinto monument on which are inscribed the names of the Texans who participated in the Battle of San Jacinto.

MOCK, William — Company D, First Regiment Texas Volunteers. He was born in Alabama in 1806 and came to Texas in January, 1835. He was married and had one son and one daughter.[1256] After the Revolution he received one league and one labor from the Harris County board.[1257] He also received a donation.[1258]

MONEY, J. H. — Company D, First Regiment Texas Volunteers. He came to Texas before 1831.[1259] He received one-fourth league in Austin's Third Colony, March 31, 1831.[1260] After the Revolution he received a donation.[1261]

MONTGOMERY, Andrew — Sixth Company, Second Regiment Texas Volunteers. Born in Alabama in 1800, he came to Texas in 1830.[1262] July 30, 1835, he received three-fourths league in Robertson's Colony.[1263] After the Revolution he received a donation.[1264]

MONTGOMERY, John — Sixth Company, Second Regiment Texas Volunteers. He was a native of Virginia.[1265] He came to Texas before February, 1835. He received one-fourth league in Robertson's Colony February 2, 1835.[1266] After the Battle of San Jacinto he received two-thirds league and one labor from the Montgomery County board.[1267] He also received a donation.[1268]

MONTGOMERY, R. W. — Company A, Regulars. He came to Texas before February, 1836 and joined the army February 23, 1836.[1269] After the Revolution he received a donation.[1270]

MOORE, Samuel — Company B, Volunteers. He arrived in Texas in 1834 and joined the army April 9, 1836. He received a bounty of 1280 acres. He also received a donation of 640 acres in Milam County.

MOORE, Joseph — Private, Company D, First Regiment Texas Volunteers. He came to Texas in 1832 and after the Revolution, he received one league and one labor of land from the Shelby County board.[1271]

MOORE, Robert—Private, Company D, First Regiment Texas Volunteers. He was born in Alabama in 1795. He came to Texas in April, 1831. He was a married man, and had one son and one daughter.[1272] He received one league in Austin's Third Colony.[1273] He also received a donation of land.[1274]

MOORE, Robert—Corporal, Company B, Volunteers. He came to Texas in January, 1836. He received one-third league of land issued by the Harris County board.[1275] After the Revolution, he received a donation of land.[1276]

MOORE, William—Corporal, Company B, Volunteers. He was a native of Tennessee and came to Texas in November 20, 1825.[1277] In December, 1835, he received ten leagues in Williams, Johnson, and Peeble's grant.[1278] After the Revolution, he received a donation of land.[1279]

MORDOFF, Henry—Company B, Volunteers. After the Revolution, he received a donation of land.[1280]

MORELAND, Isaac—Captain in the Artillery Corps. He was a native of Georgia and he came to Texas before December 26, 1834. He was a single man.[1281] In June, 1835, he received one-fourth league in Vehlein's Colony.[1282] He also received one-third league from Harris County after the Revolution.[1283] He also received a bounty of 1280 acres.[1284]

MORGAN, John—Company I, Volunteers. He came to Texas in May, 1835. After the Revolution he received one-third league of land, issued by Bastrop County.[1285] He also received a donation of land.[1286]

MORTON, John—Private, Company A, Regulars. He was a native of Kentucky and came to Texas January 21, 1826. He was a married man and had two sons and three daughters.[1287] After the Revolution, he received a bounty of 320 acres.[1288]

MOSIER, Adam—Company I, Volunteers. He was a native of Kentucky and came to Texas in October, 1835 as a member of the Volunteer Grays from New Orleans. He received a certificate for the purpose of obtaining land.[1289] After the Revolution he received a donation of land.[1290]

MOSS, John—First Company, Second Regiment Texas Volunteers. He came to Texas in 1833. He received two-thirds league and one labor, issued by Shelby County.[1291]

MOSS, Matthew—Second Company, Second Regiment Texas Volunteers. He was born in Arkansas in 1805 and came to Texas in November, 1829.[1292] He joined the army on April 1, 1836.[1293] After the Revolution he received two-thirds league and one labor of land, issued by Bastrop County.[1294] He also received a donation of land.[1295]

MOTLEY, Dr. William—Member of the Medical Staff. He was a native of Kentucky[1296] and he came to Texas before the Declaration of Independence.[1297] He was killed in the Battle of San Jacinto.[1298] His heirs received one-third of a league of land, and a donation.[1299]

MUIR, William—Company K, First Regiment Texas Volunteers. No information.

MURPHREE, David—First Lieutenant, Fourth Company, Second Regiment Texas Volunteers. He came to Texas before the Declaration of Independence. After the Revolution he received one-third of a league of land issued by Bexar County.[1300] He also received a donation.[1301]

MYERS, E. G.—Company I, Volunteers. He came to Texas before July, 1835. In July, 1835, he received one league in Burnet's Colony in Anderson County.[1302]

NABORS, William—First Company, Second Regiment Texas Volunteers. He came to Texas in February, 1836 and he received one-third league of land, issued by Fort Bend County after the Revolution.[1303] He also received a donation of land.[1304]

NASH, J. H.—Private, Cavalry Corps. He came to Texas before May 28, 1835.[1305] On July 25, 1835, he received twenty labors and 500,000 square *varas* in Zavala's Colony.[1306] After the Revolution he received a labor of land from the San Augustine County board.[1307] He also received a bounty 320 acres.[1308]

NEIL, John—Private, Cavalry Corps. He arrived in Texas in 1834. He received, after the Revolution, one league and one labor of land issued by Milam County.[1312] He also received a donation of 640 acres of land.[1313]

NEILL, J. C.—Lieutenant Colonel, Artillery Corps. He was a native of Alabama and came to Texas in February, 1831. He was a married man and had two sons and one daughter.[1309] On November 24, 1832, he received one league of land in Austin's Fourth Colony.[1310] He also received a donation of land.[1311]

NIRLAS, (First name unknown)—Private, Company B, Texas Volunteers. No information.

NELSON, David S.—Second Sergeant, Company A, Regulars. After the Revolution he received a bounty of 640 acres of land.[1314]

NELSON, James—Company F, First Regiment Texas Volunteers. He arrived in Texas in 1823.[1315] On August 7, 1824, he received one league in Austin's First Colony.[1316] After the Revolution, he received one labor of land, issued by Colorado County.[1317] He also received a donation of 640 acres of land.[1318]

NEWMAN, W. G.—Private, Company B, Volunteers. He was a resident citizen in Texas in October, 1826.[1319] After the Revolution he received one-third of a league of land issued by Liberty County.[1320] He also received a donation of 640 acres.[1321]

NIXON, N.—Private, Cavalry Corps. No information.

NOLAND, E.—Private, Company I, Volunteers. He arrived in Texas in July, 1835.[1322] After the Battle of San Jacinto, he received a donation of 640 acres of land.[1323] He also received one league and one labor of land from Harris County.[1324]

O'BANNION, Jennings—Sixth Company, Second Regiment Texas Volunteers. He came to Texas in January, 1836. He received one league of land from the Shelby County board after the Battle.[1325] He also received a donation of 640 acres of land.[1326]

O'CONNOR, P. B.—Company F, First Regiment Texas Volunteers. He came to Texas in October, 1835. He received one-third league from the Fort Bend County board after the Revolution,[1327] and a bounty of land in addition.[1328]

ODOM, David—Fifth Company, Second Regiment Texas Volunteers. He arrived in Texas in 1832. He received one-third league from the Jefferson County board after the Revolution.[1329] He also received a donation and bounty of land.[1330]

O'NEIL, John—Company A, Regulars. He was in Texas before January, 1836, and joined the army.[1331] He received a donation of 640 acres of land after the Revolution.[1332]

ORR, Thomas—Third Company, Second Regiment Texas Volunteers. He came to Texas in 1824[1333] and on May 2, 1832, he received one league and one labor of land in Liberty County.[1334] He received one-third league of land from the Liberty County board, after the Revolution.[1335] He also received a bounty of land.[1336]

OSBORNE, Benjamin—Company A, First Regiment Texas Volunteers. He was born in Mississippi and came to Texas in 1832.[1337] November 8, 1832, he received one league of land in Austin's Fourth Colony.[1338] He received a bounty of 320 acres after the Revolution.[1339]

OWNSBY, James—Company B, Volunteers. He arrived in Texas January 28, 1836. He received one-third league of land from the Harris County board, after the Battle of San Jacinto.[1340] He received in addition a bounty of 1280 acres of land.[1341]

OWEN, T. D.—Company K, First Regiment Texas Volunteers. He came to Texas before the Declaration of Independence. He received one-third of a league from the Harris County board[1342] after the Revolution. In addition he received a donation of 640 acres of land.[1343]

PACE, Dempsey—Company C, First Regiment Texas Volunteers. He arrived in Texas in 1828. He received one-third league of land from the Bastrop County board.[1344] He also received a donation of land.[1345]

PACE, James R.—Company C, First Regiment Texas Volunteers. He came to Texas in 1828. He received one-third league of land from the Bastrop County board.[1346] In addition he received a donation and a bounty of 320 acres.[1347]

PACE, Wesley—Company I, Volunteers. He arrived in Texas December 27, 1834.[1348] He came from Virginia and had a wife and eight children.[1349] December 5, 1835, he received one league of land in Vehlein's Colony in Polk County, on the east side of the Trinity.[1350] He also received a donation of land.[1351]

PACE, William—Company F, First Regiment Texas Volunteers. He came to Texas before May, 1735.[1352] May 3, 1835, he received one league of land in Vehlein's Colony in Polk County on the east side of the Trinity.[1353] He also received one-third league of land from the Colorado County board.[1354] In addition he received a donation of land.[1355]

PARK, William A.—Artillery Corps. He was a native of New York and in 1830 he came to Texas.[1356] He received one-fourth league of land in Austin's Third Colony in Matagorda County on the Tres Palacios Creek.[1357] In addition he received one-third league of land issued by the Red River County board.[1358] He also received a donation of land.[1359]

PARKER, Dickinson—First Company, Second Regiment Texas Volunteers. He arrived in Texas before May 18, 1835. He was a single man.[1360] He received three-fourths league and one labor of land issued by the Houston County board. [1361] He also received a donation of land. [1362]

PARKS, Joseph B.—Seventh Company, Second Regiment Texas Volunteers. He arrived in Texas in 1835. He received one league and one labor of land from the San Augustine County board.[1363] He also received a donation.[1364]

PASCAL, Samuel—Company B, Volunteers. He came to Texas before the Declaration of Independence. He received one-third of a league from Harris County. He also received a donation.[1365]

PATE, William—Seventh Company, Second Regiment Texas Volunteers. He came to Texas in January, 1836. He received one-third league of land from the Houston County board.[1366] He also received a donation of land.[1367]

PATTERSON, J. S.—Company I, Volunteers. He arrived in Texas before October 28, 1835.[1368] He received a donation of land.[1369]

PATTON, St. Clair—Fourth Company, Second Regiment Texas Volunteers. He arrived in Texas in May 2, 1835. He received one-third league and one labor of land from the Brazoria County board.[1370] He also received a bounty of 320 acres and a donation.[1371]

PATTON, William—Company A, Regulars. He came to Texas in 1834. He received two-thirds league and one labor from the Houston County board after the Revolution.[1372] He also received a bounty of land.[1373]

PATTON, William H.—Aid-de-camp, Second Regiment Texas Volunteers. He came to Texas in March, 1832. He was unmarried.[1374] October 10, 1835, he received one-third league of land issued by J. W. Smyth in Polk County on the Big Sandy Creek.[1375] He also received a donation of land.[1376] September, 1837, he was elected as representative from Bexar County.[1377]

PECK, Nathaniel—Company A, First Regiment Texas Volunteers. He arrived in Texas in 1832[1378] and he joined the army December, 1835.[1379] He received one league and one labor of land from the Matagorda County board.[1380] He also received a donation of land.[1381]

PECK, Nicholas—Company D, First Regiment Texas Volunteers. He arrived in Texas in 1832. He received one league and one labor of land from the Matagorda County board.[1382] He also received a donation of land.[1383]

PEEBLES, S. W.—Fifth Company, Second Regiment Texas Volunteers. He was a native of Alabama. He came to Texas, May, 1830.[1384] He was issued one-third league of land by the Brazoria County board.[1385] He also received a donation of land.[1386]

PENTECOST, G. W.—Fifth Company, Second Regiment Texas Volunteers. He arrived in Texas in January, 1828. He received one-third league of land from the Fort Bend County board.[1387] He also received a donation of land.[1388]

PERRY, James—Volunteer Aid. He was a native of Missouri. He came to Texas April 1, 1828. He was a married man and had a family of four sons and two daughters.[1389] He received one league and one labor of land from the Shelby County board.[1390] He also received a bounty of 320 acres of land.[1391]

PERRY, L. W.—Company H, First Regiment Texas Volunteers. He came to Texas before the Declaration of Independence. He received one labor of land from Montgomery County. He also received a bounty of 320 acres of land.[1392]

PETERSON, John—Sixth Company, Second Regiment Texas Volunteers. He arrived in Texas in 1822.[1393] April 2, 1831, he received one league of land in Austin's Second Colony in Grimes County.[1394] In addition he received one labor of land from the Matagorda County board.[1395] He also received a donation of land.[1396]

PETERSON, William—Sixth Company, Second Regiment Texas Volunteers. He arrived in Texas before May, 1835.[1397] July 30, 1835, he received one league of land in Robertson's Colony in Falls

County west of the Brazos River.[1398] He received one labor of land from the Montgomery County board[1399] and also a donation.[1400]

PETTUS, Edward O.—Company D, First Regiment Texas Volunteers. He arrived in Texas before the Declaration of Independence.[1401] August 12, 1835, he received ten league of land under Williams, Johnson and Peebles.[1402] He also received a donation of land.[1403]

PETTUS, John F.—Company D, First Regiment Texas Volunteers. He came to Texas in 1830[1404] and in 1831 he received one league in Austin's Second Colony.[1405] He also received one league and one labor of land from the Austin County board.[1406] He also received a donation.[1407]

PETTY, George—Company H, First Regiment Texas Volunteers. A native of Tennessee, he came to Texas in 1835.[1408] He received one league and one labor of land from the Washington County board.[1409] He also received a donation and a bounty of 320 acres.[1410]

PEVETOE, Michael—Third Company, Second Regiment Texas Volunteers. He was a native of Louisiana and came to Texas before December 10, 1834. He was married and had a family of eight children.[1411] He received one league and one labor of land from the Jefferson County board.[1412] He also received a donation and a bounty of 320 acres.[1413]

PHILIPS, Eli—Company B, Volunteers. He was born in Henrico County, Virginia, June 19, 1813. He came to Texas in August, 1835. He was a member of Captain Turner's Company at San Jacinto. He was issued one-third of a league of land February 8, 1838 by the Board of Land Commissioners for Harrisburg County, which he sold to John Belden, March 10, 1838 for $200.00. He was issued 1280 acres of land for his services, which was also sold, to James S. Holman October 25, 1837 for $200.00. He did not apply for the land due him for having participated in the Battle.

PHILIPS, Sam—First Company, Second Regiment Texas Volunteers. He arrived in Texas October 2, 1835, with his family.[1414] He joined the army March 6, 1836.[1415] He received one-third league of land from the Houston County board.[1416] He also received a bounty of 320 acres.[1417]

PHILIPS, Sidney—Fourth Company, Second Regiment Texas Volunteers. He came to Texas in May, 1834. He received one-third league of land from the Fort Bend County board.[1418] He also received a donation after the Battle of San Jacinto.[1419]

PICKERING, John—Fourth Company, Second Regiment Texas Volunteers. He came to Texas in 1820 and he received one league and one labor of land from the Jasper County board.[1420] In addition he received a donation of 640 acres.[1421]

PIERCE, (First name unknown)—Company A, Regulars. No information.

PIERCE, W. J. C.—Cavalry Corps. He arrived in Texas in January, 1835. He received one league and one labor of land from the Fort Bend County board.[1422] He also received a donation and a bounty of 320 acres.[1423]

PINCHBACK, James R.—Company A, First Regiment Texas Volunteers. He came to Texas before the Declaration of Independence.[1424] He joined the army February 15, 1836.[1425] After the Battle he received one-third league from the Gonzales County board.[1426] Also he received a donation and a bounty of 1280 acres.[1427]

PLASTER, Thomas P.—Artillery Corps. He came to Texas February, 1836. He received one league and one labor from the Harris County board.[1428] He also received a donation and a bounty of 320 acres.[1429]

PLUNKETT, J.—Company K, First Regiment Texas Volunteers. He came to Texas in 1834 and he received one-third league of land from the Matagorda County board.[1430] He also received a bounty of 320 acres and a donation.[1431]

POWELL, James—Company H, First Regiment Texas Volunteers. He was a native of Alabama and came to Texas in 1832.[1432] February 22, 1836, he received one-fourth league of land in Austin's Fifth Colony in Grimes County.[1433] After the Battle of San Jacinto, he received one-third league from the Fort Bend County board.[1434] In addition he received a donation.[1435]

PRATT, Thomas—Company I, Volunteers. He arrived in Texas in 1835. He received one-third league of land from the Harris County board.[1436] He also received a donation.[1437]

PRUITT, Levi—First Company, Second Regiment Texas Volunteers. He came to Texas before the Declaration of Independence. He received one league and one labor of land from Houston County. In addition he received a donation.[1438]

PROCTOR, J. W.—Eighth Company, Second Regiment Texas Volunteers. He came to Texas in December, 1835. He received one-third league from the Robertson County board.[1439] He also received a bounty of 640 acres and a donation.[1440]

PUTMAN, Mitchell—Company F, First Regiment Texas Volunteers. He came to Texas in 1835. He received one league and one labor from the Jackson County board after the Revolution.[1441] He also received a donation of land.[1442]

RAINWATER, E. R.—Cavalry Corps. He came to Texas before the Declaration of Independence. After the Revolution he received one-third league from the Montgomery County board.[1443]

RAMEY, Lawrence—Sixth Company, Second Regiment Texas Volunteers. He came to Texas before May, 1827. He received one league in Austin's First Colony.[1444] After the Revolution he received a donation.[1445]

RAMIREZ, Eduardo—Ninth Company, Second Regiment Texas Volunteers. He was a native of Bexar and after the Battle of San Jacinto he received one league and one labor from the Bexar County board.[1446] He received a donation also.[1447]

RAINEY, Clement—Company H, First Regiment Texas Volunteers. He came to Texas before March, 1835. He received one league of land in Robertson's Colony on March 18, 1835.[1448] After the Battle of San Jacinto he received a bounty of 640 acres.[1449]

RAYMOND, Samuel B.—Second Lieutenant, Company A, First Regiment Texas Volunteers. He was a native of Tennessee and came to Texas in 1831. He had a wife and four children.[1450] After the Revolution he received a donation.[1451]

RECTOR, Claiborn—Fourth Company, Second Regiment Texas Volunteers. He was born in Alabama in 1803 and he came to Texas in February, 1830. He was not married and was a farmer.[1452] After the Revolution he received one league and one labor from the Fort Bend County board.[1453] He also received a donation. [1454]

RECTOR, E. G.—Fourth Company, Second Regiment Texas Volunteers. He came to Texas before December, 1835. After the Revolution he received one-third league from the Fort Bend County board.[1455] He also received a donation.[1456]

RECTOR, Pendleton—Fourth Company, Second Regiment Texas Volunteers. He was born in Alabama in 1805 and he came to Texas in February, 1830. He was not married. He was a farmer.[1457] After the Revolution he received one-third league from the Brazoria County board.[1458] He also received a donation.[1459]

REDD, William D.—Cavalry Corps. He came to Texas after the Declaration of Independence. After the Revolution he received one-third league in Harris County. He also received a bounty of 1280 acres.[1460] He was killed in a duel by Major Lysander Wells in San Antonio in 1840.[1461]

REED, Henry—Eighth Company, Second Regiment Texas Volunteers. He came to Texas before May, 1835.[1462] He received one-fourth league in Burnet's Colony.[1463] After the Revolution he received one labor from the Robertson County board.[1464] He also received a donation.[1465]

REEL, R. J. W.—Company I, Volunteers. He came to Texas before May 2, 1835.[1466] He received one league of land in Austin's Fifth

Colony.[1467] After the Revolution he received a donation.[1468] He also received one labor from the Brazoria County board.[1469]

REESE, C. K.—Company K, First Regiment Texas Volunteers. He was born in Tennessee in 1801. He came to Texas in February, 1830. He was not married. He was a farmer.[1470] June 20, 1831, he received one league in Austin's Third Colony.[1471] After the Revolution he received one-twelfth league from the Brazoria County board.[1472] He also received a donation.[1473] He went on the Mier Expedition in 1842.[1474]

REESE, W. P.—Company K, First Regiment Texas Volunteers. He came to Texas before May 2, 1835. After the Revolution he received one labor from the Brazoria County board.[1475] He also received a donation.[1476]

REEVES, D. W.—Cavalry Corps. He came to Texas in June, 1835. After the Battle of San Jacinto he received one-third league from the Nacogdoches County board.[1477]

REID, Nathaniel—Company F, First Regiment Texas Volunteers. He came to Texas before the Declaration of Independence. After the Revolution he received one league and one labor from the Austin County board.[1478] He also received a donation.[1479]

RHEINHART, J. P.—Company A, Regulars. He came to Texas before the Declaration of Independence. After the Revolution he received one-third league from the Nacogdoches County board. He also received a donation.[1480]

RHODES, Joseph—Company A, First Regiment Texas Volunteers. Born in Mississippi, he arrived at Matagorda, Texas, January 10, 1836 as a member of Captain Vickery's Company. He also joined Captain Sherman's Company, of which he was elected First Lieutenant, March 13, 1836, serving as such in Captain Wood's Company at San Jacinto. Lieutenant Rhodes mortally wounded Sergeant John Pollard in a duel March 24, 1837, at Houston. Pollard died March 31st. Rhodes did not apply for headright, donation or bounty certificates.

RIAL, John W.—Company A, First Regiment Texas Volunteers. He came to Texas in February, 1836. After the Revolution he received one labor from the Nacogdoches County board.[1481] He also received a donation.[1482]

RICHARDSON, (First name unknown)—Private, Company A, Regulars. No information.

RICHARDSON, (First name unknown)—First Corporal, Company A. No information.

RICHARDSON, John—Sixth Company, Second Regiment Texas Volunteers. He arrived in Texas in 1834 and after the Revolution he received one-third league from the Harris County board.[1483] He also received a donation.[1484]

RICHARDSON, William—Company A, Volunteers. He came to Texas before September, 1835.[1485] He received one-fourth league of land in Vehlein's Colony.[1486] After the Revolution he received one-third league in Jasper County.[1487]

RIPLEY, Phineas—Fourth Company, Second Regiment Texas Volunteers. He came to Texas before May 2, 1835. After the Battle he received one-third league from the Brazoria County board.[1488] He also received a bounty of 320 acres.[1489]

ROBBINS, John—He was a native of Arkansas and came to Texas in 1825. He was married and had two daughters.[1490] After the Revolution he received one league and one labor from the Red River County board.[1491] He also received a donation.[1492]

ROBBINS, Thomas—Cavalry Corps. He came to Texas before July 17, 1835.[1493] He received a donation after the Battle of San Jacinto.[1494]

ROBERTS, David—Seventh Company, Second Regiment Texas Volunteers. He arrived in Texas before the Declaration of Independence. After the Revolution he received one-third league in Houston County. He also received a donation.[1495]

ROBERTS, Sim—Seventh Company, Second Regiment Texas Volunteers. No information.

ROBERTSON, William—Third Company, Second Regiment Texas Volunteers. He came to Texas in 1831 and received one league and one labor from the Sabine County board after the Revolution.[1496]

ROBINSON, George W.—Second Company, Second Regiment Texas Volunteers. He came to Texas before June 15, 1834. He was married and had one child.[1497] After the Battle he received two-thirds league and one labor from the Montgomery County board.[1498] He also received a donation.[1499]

ROBINSON, James W.—Cavalry. A native of Ohio, he was a member of the Consultation in 1835 from Nacogdoches. He was appointed Lieutenant Governor at the organization of the provisional government. He was appointed a District Judge when the Constitutional Government was organized. He was taken prisoner by General Woll in 1842, when Mexico invaded Texas. He was released by Santa Anna. In 1849 Judge Robinson removed with his family to California. He died there at San Diego in 1853.[1500]

ROBINSON, Joel W.—Company F, First Regiment Texas Volunteers. He was born in Washington County, Georgia in 1815. He came to Texas in 1831 and landed at the mouth of the Brazos. He was in the Velasco fight in 1832. Mr. Robinson was with the party that captured Santa Anna.[1501] He received a donation of land after the Revolution.[1502]

ROBINSON, T. J.—Artillery Corps. He was born in Alabama in 1813. He came to Texas in February, 1835. He was a blacksmith and was not married.[1503] After the Revolution he received one-third league from the Montgomery County board.[1504]

ROCKWELL, C.—Seventh Company, Second Regiment Texas Volunteers. He came to Texas before May 2, 1835 and after the Revolution he received one-third league from the San Augustine County board.[1505]

RODRIGUEZ, Ambrosio—Ninth Company, Second Regiment Texas Volunteers. He was a native of Bexar and after the Battle of San Jacinto he received one league and one labor from the Bexar County board.[1506] He also received a donation.[1507]

ROLLISON, (First name unknown)—Sixth Company, Second Regiment, Texas Volunteers. No information.

ROMAN, Richard—Captain, Company B, Volunteers. He was a native of Kentucky[1508] and came to Texas before the Declaration of Independence. He received one-third league from the Goliad board after the Revolution. He also received a donation.[1509] He was a Major in the Mexican War in 1846. In 1836 he represented Victoria in the first Congress. He later went to California where he served as Treasurer of the State.[1510] He was appointed Appraiser of Merchandise in San Francisco. He died there in 1876.[1511]

RORDER, Louis—Company D, First Regiment Texas Volunteers. No information.

ROUNDS, Lyman F.—Company C, Regulars. He came to Texas in 1835.[1512] He joined the army January 19, 1836.[1513] After the Revolution he received one-third league from the San Augustine County board.[1514] He received a bounty of 1280 acres.[1515]

ROWE, James—First Lieutenant, Eighth Company, Second Regiment Texas Volunteers. He came to Texas before December 3, 1834. He was not married.[1516]

RUDDER, Nathaniel—Company I, Volunteers. He came to Texas before the Declaration of Independence. After the Revolution he received one-third league from the Brazoria County board.[1517] He also received a donation.[1518]

RUDDLE, John—Third Company, Second Regiment Texas Volunteers. He came to Texas before the Declaration of Independence. After the Revolution he received one-third league from the Liberty County board. He also received a donation.[1519]

RUSK, David—First Company, Second Regiment Texas Volunteers. He came to Texas in January, 1836.[1520] January 30, 1836, he received ten leagues from Durst and Williams.[1521] After the Revolution he received one-third league from the Nacogdoches County board.[1522] He also received a donation.[1523]

RUSK, Thomas J.—Secretary of War. He was born in South Carolina in 1803. John C. Calhoun aided him with his education. After he became a lawyer, he moved to Georgia, where he obtained a successful practice. He came to Texas in 1834, and at once took prominent part in the affairs of the people. In 1835 the Executive Council elected him Commissary of the Army. He signed the Declaration of Independence in March, 1836. At the organization of the Government *ad interim*, he entered Burnet's Cabinet as Secretary of War. He performed a gallant part in the Battle of San Jacinto. When General Houston went to New Orleans for surgical aid, Rusk was made Commander-in-Chief of the Army. In the fall of 1836 Rusk was appointed Secretary of War, but resigned soon afterwards. In 1837 he was a member of the Texas Congress. In 1843 he was elected Major-General of the Militia. In 1845 he was elected President of the Annexation Convention. He was sent to the United States Senate after Annexation, a position he held until his death in 1857.[1524]

RUSSELL, R. B.—Seventh Company, Second Regiment Texas Volunteers. He received one league in Vehlein's Colony on December 23, 1834.[1525] After the Revolution he received one-twelfth league from the San Augustine County board.[1526] He also received a bounty of 320 acres.[1527]

RYONS, Thomas—Company F, First Regiment Texas Volunteers. He received a bounty of 320 acres.[1528]

SADLER, John—Second Company, Second Regiment Texas Volunteers. He was a native of Tennessee and came to Texas before 1834. He had a wife and two children.[1529] April 29, 1835, he received one league in Vehlein's Colony.[1530] After the Revolution he received one-third league of land from the Montgomery County board.[1531] He also received a donation.[1532]

SADLER, William T.—First Company, Second Regiment Texas Volunteers. He arrived in Texas in 1835 and received one league and one labor from the Houston County board after the Revolution.[1533] He also received a donation.[1534]

SANDERS, Uriah—Company H, First Regiment Texas Volunteers. He came to Texas before November 9, 1835 and received one league of land in Caldwell County.[1535] He received one league and one labor from the Austin County board after the Revolution.[1536] He also received a donation of land.[1537]

SAUNDER, John—Company A, Regulars. He arrived in Texas January 1, 1834. He had a wife and one child.[1538] He joined the army January 8, 1836.[1539] After the Revolution he received one-third league from the Red River County board,[1540] and he received a donation in addition.[1541]

SAYRES, John—Sixth Company, Second Regiment Texas Volunteers. He was born in Kentucky and came to Texas in February, 1836.[1542] He received a donation of land after the Revolution.[1543]

SCALLIONS, J. W.—Sixth Company, Second Regiment Texas Volunteers. He came to Texas in December, 1835. He received one-third league from the Fayette County board.[1544] He also received a donation after the Revolution.[1545]

SCALLOM, (First name unknown)—Sixth Company, Second Regiment Texas Volunteers. No information.

SCARBROUGH, Paul—Company D, First Regiment Texas Volunteers. He arrived in Texas in 1830. He received one-third league of land from the Jackson County board.[1546] He also received a donation of land.[1547] He went on the Santa Fe Expedition in 1842.[1548]

SCATES, William B.—Seventh Company, Second Regiment Texas Volunteers. He arrived in Texas March 2, 1832. He received one league and one labor from the Washington County board after the Revolution.[1549] He also received a donation of land.[1550]

SCHESTON, Henry—Company B, Volunteers. He received a bounty of land after the Revolution.[1551]

SCOTT, W. P.—Company K, First Regiment Texas Volunteers. He arrived in Texas in 1830 and he received one league and one labor of land from the Sabine County board after the Revolution.[1552] He also received a donation of land.[1553]

SCURRY, Richardson—First Sergeant, Artillery Corps. He arrived in Texas before the Declaration of Independence. He received one-third league from the San Augustine County board after the Revolution.[1554] He also received a donation[1555] and on December 16, 1836, he was elected Prosecuting Attorney for the First District.[1556]

SECREST, Fielding—Cavalry Corps. He came to Texas before the Declaration of Independence. He received two-thirds league and one labor from the Harris County board after the Revolution.[1557] He also received a donation of land.[1558]

SECREST, Washington—Cavalry Corps. He came to Texas before May 2, 1835. He received one league and one labor from the Colorado County board.[1559] He also received a donation.[1560]

SEGUIN, Juan N.—Captain, Ninth Company, Second Regiment Texas Volunteers. He was Political Chief of the Department of Bexar before the Revolution in 1834. Although a Mexican, he espoused the cause of Texas and joined them in resisting the advance of Santa Anna. After the retreat of the Mexicans, he was promoted to the rank of Colonel and appointed commander of his native city. In 1839 he represented Bexar County in the Senate. Later he left the country because he had serious personal misunderstandings with some of the Americans of San Antonio. When

Woll invaded Texas in 1842, Seguin was one of his staff officers, and fought against the Texans in the Battle of Salado. He was a Colonel in the Mexican army at the Battle of Buena Vista. He resigned soon afterwards and made his way back to Texas with his family.[1561] He received a donation of land after the Battle of San Jacinto.[1562]

SELF, George—Company C, First Regiment Texas Volunteers. He came to Texas before the Declaration of Independence. He received a donation after the Battle. He also received a bounty of 640 acres and one-third league in Bastrop County.[1563]

SENNETT, Andrew—Company F, First Regiment Texas Volunteers. No information.

SERGEANT, W.—Company I, Volunteers. He received a donation after the Battle.[1564]

SEVEY, Manasseh—Company A, First Regiment Texas Volunteers. He came to Texas in January, 1836. After the Revolution he received one-third league in Harris County.[1565] He also received a donation.[1566]

SEVEY, Ralph E.—Company A, First Regiment Texas Volunteers. He came to Texas in 1836 and after the Revolution he received one-third league in Harris County.[1567] He also received a donation.[1568]

SHARPE, John—First Lieutenant, First Regiment Texas Volunteers. He came to Texas before the Declaration of Independence. After the Revolution he received a donation and a bounty of 320 acres. He also received one league and one labor in Brazoria County.[1569]

SHAW, James—Cavalry Corps. He came to Texas before 1832. On June 16, 1832, he received one-fourth league in DeWitt's Colony.[1570] After the Revolution he received one league and one labor in Brazoria County.[1571] He also received a donation.[1572] He represented Milam County in the Second Legislature in 1838.[1573]

SHERMAN, Sidney—Colonel, Second Regiment Texas Volunteers. He was born in Massachusetts on July 23, 1805. He came to Texas in January, 1836.[1574] After the Revolution he received one league and one labor from the Harris County board.[1575] He also received a donation.[1576] He died on August 1, 1873.[1577]

SHREVE, John M.—Company I, Volunteers. He was born in Kentucky in 1811. He came to Texas in July, 1835. He was a clerk.[1578] After the Battle he received one league and one labor from the Harris County board.[1579] He also received a bounty of 320 acres.[1580] He was chief clerk of the called session of the Legislature September 26, 1837.[1581]

SHUPE, Samuel—Fifth Company, Second Regiment Texas Volunteers. He was a native of Pennsylvania.[1582] May 5, 1831, he received one-fourth league in DeWitt's Colony.[1583] After the Battle of San Jacinto he received two-thirds league and one labor from the Jackson County board.[1584] He also received a donation.[1585]

SIGMAN, Abel—Company B, Volunteers. He came to Texas before the Declaration of Independence. He received one-third league in Fort Bend County after the Revolution. He also received a bounty.[1586]

SIMMONS, William—Company C, First Regiment Texas Volunteers. He came to Texas in April, 1835. He received one-third league from the Bastrop County board.[1587] He also received a bounty.[1588]

SIMPSON, J.—Company H, First Regiment Texas Volunteers. He came to Texas in 1834[1589] and he received eleven leagues from James Bowie in Anderson County, on the waters of the Trinity.[1590] After the Revolution he received a donation.[1591]

SLACK, Joseph H.—Company I, Volunteers. He came to Texas in November, 1835. After the Revolution he received one-third league from the Fort Bend County board.[1592] He also received a donation.[1593]

SLAYTON, John—Third Company, Second Regiment Texas Volunteers. He came to Texas in 1825. After the Revolution he received one-third league in Jasper County.[1594]

SMITH, (First name unknown)—Company A, Regulars. No information.

SMITH, (First name unknown)—Company B, Volunteers. No information.

SMITH, (First name unknown)—Private, Company A, Regulars. No information.

SMITH, (First name unknown)—Private, Company B, Volunteers. No information.

SMITH, Benjamin F.—Cavalry Corps. He was a native of Kentucky and represented Hines County in the First Legislature of Mississippi.[1595] He came to Texas in January, 1836. After the Revolution he received one-third league from the Harris County board.[1596] He also received a donation.[1597] He was in the Texas Congress in 1840. He died in Montgomery County in 1841.[1598]

SMITH, Erastus (Deaf)—Cavalry Corps. He was born in New York and he came to Texas in May, 1821. He was a spy and a guide for the Texans during the Revolution.[1599] On October 5, 1835, he received one-fourth league of land in Robertson's Colony.[1600] After the Revolution he received one league and one labor from the Brazoria County board.[1601] In 1837 he was captain of a ranging force in the west. Later he retired to civil life and lived in Richmond. He and John P. Borden established a land agency business together. He died in Richmond, November 30, 1837.[1602]

SMITH, Hugh M.—Artillery Corps. He came to Texas in 1832. After the Revolution he received one league and one labor from Red River County.[1603]

SMITH, James—Company K, First Regiment Texas Volunteers. He came to Texas in April, 1835. After the Revolution he received one league and one labor from the Nacogdoches County board.[1604] He also received a donation.[1605]

SMITH, John—Sergeant Major, Staff of the Command. He came to Texas March 10, 1831.[1606] February 26, 1835 he received twenty leagues in Zavala's Colony.[1607] He received one-third league from the Montgomery County board after the Revolution.[1608] He also received a donation.[1609]

SMITH, John—Fourth Company, Second Regiment Texas Volunteers. He was a native of Ohio and came to Texas in January, 1830. He was married and had two sons and two daughters. He was a farmer.[1610] After the Revolution he received one league and one labor from the Brazoria County board.[1611] He also received a donation.[1612]

SMITH, R. W.—First Company, Second Regiment Texas Volunteers. He came to Texas in March, 1835.[1613] He joined the army on March 6, 1836.[1614] May 13, 1835, he received one-fourth league in Burnet's Colony.[1615] After the Revolution he received three-fourths league and one labor from the Nacogdoches County board.[1616] He also received a donation.[1617]

SMITH, W. A.—Third Company, Second Regiment Texas Volunteers. He came to Texas in 1831. He received one-third league from Liberty County board.[1618] After the Revolution he received a bounty of 320 acres.[1619]

SMITH, William—Third Company, Second Regiment Texas Volunteers. He was born in Georgia in 1802. He came to Texas in August, 1830. He was farmer.[1620] After the Revolution he received one league and one labor from the Shelby County board.[1621] He also received a donation.[1622]

SMITH, William H.—Captain, Cavalry Corps. He came to Texas in 1829.[1623] December 12, 1835, he received ten leagues from Williams, Johnson and Peebles.[1624] After the Revolution he received one league and one labor from the Jefferson County board.[1625] He also received a donation.[1626]

SMITH, William M.—Third Company, Second Regiment Texas Volunteers. He came to Texas in 1827. After the Revolution he received one league and one labor from the Liberty County board.[1627] He also received a donation.[1628]

SNELL, Martin K.—First Lieutenant, Company A, Regulars. He was born in Pennsylvania.[1629] He came to Texas in 1835.[1630] After the Revolution he received one-third league from the Harris County board.[1631] He also received a donation.[1632]

SNYDER, M.—Corporal, Company B, Volunteers. He came to Texas in 1831. He received one-third league from the Harris County board after the Battle of San Jacinto.[1633]

SOMERVELL, Alexander—Lieutenant-Colonel, First Regiment Texas Volunteers. He came to Texas in 1832.[1634] He received one-fourth league in Austin's Second Colony on April 29, 1832.[1635] He was elected Mayor, at the organization of the army, in Burleson's army. When the army was reorganized he was made Lieutenant Colonel. In Burnet's Cabinet he was Acting Secretary of War, a posted vacated by Lamar. He was afterwards a Senator in the First Congress.[1636] After the Revolution he received one-twelfth league from the Austin County board.[1637] He also received a donation.[1638] There has never been a satisfactory explanation of the manner of his death. He started from Saluria to Lavaca in January, 1854, with some money. He was found lashed to the timbers of his over-turned boat. It is not known whether he was murdered or whether his death was accidental.[1639]

SOVEREIGN, Joseph—Company I, Volunteers. He came to Texas before the Declaration of Independence. After the Revolution he received one-third league from the Brazoria County board.[1640] He also received a donation.[1641]

SPARKS, S. F.—First Company, Second Regiment Texas Volunteers. He came to Texas in January, 1835. After the Revolution he received one league and one labor from the Nacogdoches County board.[1642] He also received a donation.[1643]

SPICER, J. A.—Company K, First Regiment Texas Volunteers. He came to Texas before the Declaration of Independence. He received one-third league in Matagorda County after the Battle of San Jacinto. He also received a bounty of 320 acres.[1644]

STANDEFER, William—Company C, First Regiment Texas Volunteers. He came to Texas in 1824. After the Revolution he received one-third league from the Bastrop County board.[1645] He also received a donation.[1646]

STANDIFORD, Jacob—Company C, First Regiment Texas Volunteers. He came to Texas in 1829 he received one-third league from the Bastrop County board after the Revolution.[1647] He also received a donation.[1648]

STARKLEY, B. F.—Company I, Volunteers. No information.

STEBBINS, Charles—Company I, Volunteers. He arrived in Texas in 1836. He joined Major Smith's Company, April 5, 1836. He was a member of Captain William S. Fisher's Company of Velasco Blues at San Jacinto. On September 8, 1838 he was issued 640 acres of land for having participated in the Battle, and 320 acres for having served in the army. The land office records do not show that he applied for a headright Certificate. He died in Collin County in 1880.

STEELE, Alphonzo—Sixth Company, Second Regiment Texas Volunteers. He came to Texas in June 1836. After the Revolution he received one league from the Montgomery County board.[1649] He also received a donation.[1650]

STEELE, Maxwell—Company F, First Regiment Texas Volunteers. He came to Texas before May 2, 1835. After the Revolution he received one-third league from the Colorado County board.[1651] He also received a donation. [1652]

STEVENSON, Robert—Company H, First Regiment Texas Volunteers. He was a native of Tennessee. He was married and had one son and one daughter.[1653] After the Battle of San Jacinto he received a donation.[1654]

STEVENS, Ashley R.—Company H, First Regiment Texas Volunteers. He was born in Tennessee in 1806. He came to Texas in February, 1831. He was a farmer.[1655] November 22, 1832, he received one league in Austin's Second Colony.[1656] He was killed in the Battle of San Jacinto.[1657] His heirs received a donation after the Revolution.[1658]

STEVENSON, Robert—Commanding Company H, First Regiment Texas Volunteers. (Name given only in the Muster Rolls at the Land Office.) He was born in Tennessee.[1659] He came to Texas in 1832.[1660] He was married and had a family.[1661] He received one league in Austin's Second Colony on November 19, 1832.[1662] After the Revolution he received one labor from the Washington County board.[1663] Additionally, he received a donation.[1664]

STEVENSON, R.—Eighth Company, Second Regiment Texas Volunteers. He came to Texas in 1834. After the Battle of San Jacinto he received one league and one labor from the Sabine County board.[1665]

STEWART, Charles—Company B, Volunteers. He came to Texas before 1821. He received one-fourth league in Austin's Second Colony on May 17, 1831.[1666] After the Revolution he received one league and one labor from the Brazoria County board.[1667]

STILWELL, W. S.—First Lieutenant, Artillery Corps. He received a bounty of 1285 acres and a donation after the Revolution.[1668]

STONFER, Henry S.—Company D, First Regiment Texas Volunteers. He received a bounty of 1290 acres after Revolution.[1669]

STOWE, Phillip—Wyly's Company. (Name given only in Muster Rolls at the Land Office.) No information.

STROTH, Philip—Company D, First Regiment Texas Volunteers. He received a donation after the Battle of San Jacinto.[1670]

STROUD, John W.—Company I, Volunteers. He was born in Virginia in 1813. He was a member of Captain Fisher's Company. He was not issued headright, bounty or donation Certificate. He was issued 320 acres of land, April 18, 1838 for having served in the army and an additional 320 acres of land September 30, 1841.

STUMP, J. S.—Company H, First Regiment Texas Volunteers. He arrived in Texas in June, 1835. After the Revolution he received one-third league from the Gonzales County board.[1671] He also received a donation.[1672] He participated in the Santa Fe Expedition in 1842.[1673]

SULLIVAN, Dennis—Company A, Regulars. He came to Texas in 1835.[1674] He joined the army on January 5, 1836.[1675] After the Revolution he received a bounty.[1676] He also received one league and one labor from the Shelby County board.[1677]

SUMMERS, W. W.—Company B, Volunteers. He came to Texas in January, 1830. After the Battle of San Jacinto he received three-fourths league and one labor from the Harris County board.[1678] He also received a donation.[1679]

SUTHERLAND, George—Company D, First Regiment Texas Volunteers. He arrived in Texas in 1830 from Alabama.[1680] He received one league in Austin's Third Colony November 24, 1830.[1681] He received one labor from the Jackson County board.[1683] He also received a donation.[1684] He was in the Convention in 1833.[1682] He was a member of the Second Congress of the Republic.[1685] He was Major in the army in 1842 to repel the Mexican Invasion.[1686] He died in Jackson County in 1855.[1687]

SWAIN, William T.—Company A, Regulars. He came to Texas in 1834.[1688] He received one-fourth league in Robertson's Colony, March 14, 1835.[1689] After the Battle he received a donation.[1690]

SWEARINGEN, N.—Company B, Volunteers. No information.

SWEARINGEN, V. W.—Company D, First Regiment Texas Volunteers. He came to Texas in March, 1835. After the Revolution he received one-third league in Austin County.[1691] He also received a donation.[1692]

SWEENY, Thomas—Fourth Company, Second Regiment Texas Volunteers. He arrived in Texas before May 2, 1835. After the Battle of San Jacinto he received one-third league from the Brazoria County board.[1693] He also received a donation.[1694]

SWEENY, W. B.—Cavalry Corps. He came to Texas in January 20, 1836. February 16, 1836, he received one league in Austin's Fifth Colony.[1695] After the Revolution he received a donation.[1696]

SWISHER, H. H.—First Lieutenant, Company H, First Regiment Texas Volunteers. He came to Texas before May, 1830. After the Revolution he received one league and one labor from the Washington County board.[1697] He also received a donation.[1698]

SWISHER, John M.—Company H, First Regiment Texas Volunteers. He arrived in Texas in 1833. After the Revolution he received one-third league from the Washington County board.[1699] He also received a donation.[1700] He was elected State Auditor in 1849 and was re-elected in 1851.[1701]

SYLVESTER, James A.—Company A, First Regiment Texas Volunteers. He was born in Baltimore, Maryland.[1702] He arrived in Texas in February, 1836. After the Battle of San Jacinto he received one-third league from the Jackson County board.[1703] He also received a donation.[1704]

TANCLE, S.—Company I, Volunteers. No information.

TANNER, E. M.—Third Company, Second Regiment Texas Volunteers. He arrived in Texas previous to the Declaration of Independence. He received one-third league of land from the Liberty County board.[1705] In addition he received a donation of land.[1706]

TARIN, Manuel—Ninth Company, Second Regiment Texas Volunteers. He was a native of Bexar. He received one league and one labor from the Bexar County board.[1707] He also received a donation of land.[1708]

TARLTON, James—Company D, First Regiment Texas Volunteers. He came to Texas from Kentucky in December, 1835.[1709]

TAYLOR, (First name unknown)—Company A, Regulars. No information.

TAYLOR, Abraham—Corporal, Company B, Volunteers. He arrived in Texas, January 28, 1836. He was recruited in New Orleans by Captain Turner and landed at Velasco January 28th. He was First Corporal in Captain Roman's Company at San Jacinto. He received one-third of a league of land by the Montgomery County Board of Land Commissioners. On June 11, 1838 he was issued 640 acres of land for having participated in the Battle. On March 4, 1839 he was issued 640 acres of land for having served in the army.

TAYLOR, Edward W.—Company A, First Regiment Texas Volunteers. He arrived in Texas previous to the Declaration of Independence.[1710] In December, 1835, he joined the army.[1711] After the Battle, he received one-third league of land from the Liberty County board.[1712] He also received a bounty of 960 acres of land.[1713]

TAYLOR, J. B.—Fourth Company, Second Regiment Texas Volunteers. He came to Texas from New York in 1830 and was a carpenter, married with one son.[1714] He received one-third league of land from the Fort Bend County board.[1715]

TAYLOR, William S.—Cavalry. He came to Texas before August, 1831. On August 8, 1831 he received one league of land in Austin's Colony in Lavaca County.[1716] After the Battle he received one-third league of land from the Milam County board. He also received a donation.[1717]

THOMAS, Benjamin—Eighth Company, Second Regiment Texas Volunteers. He was a native of Tennessee and was born in 1818. He joined the army January 14, 1836.[1718] He came to Texas February, 1835. He received one-third league of land from the Austin County board.[1719] He also received a donation of land.[1720]

THOMPSON, (First name unknown)—Company I, Volunteers. No information.

THOMPSON, Charles P.—Company F, First Regiment Texas Volunteers. Arrived in Texas in 1825. He received one league and one labor from the Shelby County board.[1721] He also received a donation.[1722]

THOMPSON, Cyrus W.—Third Company, Second Regiment Texas Volunteers. He arrived in Texas before 1835. He came from New York and was unmarried.[1723] He received one-fourth league of land from the Liberty County board.[1724] He also received a donation of land.[1725]

THOMPSON, Jesse—Second Company, Second Regiment Texas Volunteers. He arrived in Texas January, 1824[1726] and joined the army April 3, 1836.[1727] He received on August 7, 1824, one league of land in Austin's First Colony in Brazoria County.[1728] After the Battle he received one-third league of land from the Fort Bend County board.[1729] He also received a donation of land.[1730]

TIERWESTER, Henry—Company I, Volunteers. He was a native of Ohio, born in 1795. In August, 1828, he came to Texas.[1731] October 19, 1832, he received one-fourth league of land in Austin's Second Colony in Harris County.[1732] After the Battle he received two-thirds of a league and one labor from the Harris County board.[1733] He also received a donation of land.[1734]

TINDALL, D.—Company B, Volunteers. He came to Texas in January, 1836. He received one-third league of land from Harris County.[1735]

TINDALL, William M.—Company A, Regulars. He arrived in Texas before January 15, 1836. He joined the army January 15,

1836.[1736] He received one-third league of land from the Harris County board.[1737] He also received a bounty.[1738]

TINSLEY, James W.—Adjutant, First Regiment Texas Volunteers. He arrived in Texas before the Declaration of Independence. He received one-third league of land from the Bexar County board.[1739] He also received a donation of land.[1740]

TONG, John B.—Company H, First Regiment Texas Volunteers. He arrived in Texas in March, 1827.[1741] He received one-fourth league of land in Montgomery County.[1742] He also received three-fourths league and one labor issued by Montgomery County after the Battle of San Jacinto.[1743]

TOWNSEND, Spencer—Cavalry. He arrived in Texas before the Declaration of Independence. He received one-third league of land from the Colorado County board.[1744] He also received a donation of land.[1745]

TREADGILL, Joshua—Company K, First Regiment Texas Volunteers. He came to Texas in 1836. He received one-third league of land from the Matagorda County board.[1746] He also received a donation of land.[1747]

TRENARY, John B.—First Company, Second Regiment Texas Volunteers. He arrived in Texas before 1835. He was a married man and he owned two slaves.[1748] He received a donation of land.[1749]

TUMLINSON, John—Company F, First Regiment Texas Volunteers. He was a native of Tennessee. He came to Texas in November, 1821. He was a married man and he had three sons. He was a farmer by occupation.[1750] He received one league and one labor of land from the San Augustine County board.[1751] He also received a donation of land.[1752]

TURNAGE, S. C.—Cavalry Corps. He arrived in Texas before May, 1835. He received one league and one labor of land from the Washington County board.[1753] He also received a donation of land.[1754]

TURNER, Amasa—Captain, Company B, Volunteers. He was born in Massachusetts in 1800. He came to Texas in 1835, and settled in Bastrop. He was the first to receive a Captain's Commission from General Houston after his appointment as Commander of the Army in 1835.[1755] He received one league and one labor from the Harris County board after the Revolution,[1756] and also a donation of 640 acres.[1757] After the Battle, he was for a time commander of the post of Galveston. While at Velasco on business in June, 1836, he thwarted the attempt to arrest President Burnet and overthrow the Civil Government. During the Republic, Colonel Turner resided at Galveston, having been one of the first to settle upon the island. He served in the Legislature in 1850 and 1851, and in the Senate in 1852-53. During the Civil War, he was Provost-Marshal of Lavaca County. At the close of the war he moved to Gonzales where he died July 21, 1877.[1758]

TYLER, Charles C.—Company B, Volunteers. He arrived at Velasco January 28, 1836, having been recruited by Captain Turner. He was a member of Captain Turner's Company at San Jacinto. He was transferred to the Navy April 8, 1837. He was issued one-third of a league of land by the Harrisburg County Board on June 7, 1838. On November 27, 1837, he was issued 960 acres of land for having served in the army. He sold it November 27, 1837, for $50.00 to Jeremiah Sanders. July 23, 1839, he was issued 640 acres of land for having participated in the Battle.

TYLER, Robert D. T.—Fifth Company, Second Regiment Texas Volunteers. He arrived in Texas before May 2, 1835. He received one-third league of land from the Brazoria County board.[1759] He also received a donation.[1760]

USHER, Patrick—Private in Company D, First Regiment Texas Volunteers. He was a native of Ireland[1761] and came to Texas in August, 1834. After the Revolution he received one-third league of land, issued by Jackson County.[1762] He also received a bounty of land 320 acres.[1763] He was elected County Judge of Jackson County, December 16, 1836,[1764] and was a member if the Fifth

Congress which met in November, 1840.[1765] He went on the Mier Expedition in 1842, was captured and died in the prison of Perote[1766] in 1843.

VAN WINKLE, John—Private, Company A, Regulars. He came to Texas in January, 1835.[1767] He joined the army January 22, 1836.[1768] After the Revolution, he received a donation of 640 acres.[1769]

VANDEVER, Logan—Private, Company C, First Regiment Texas Volunteers. He was born in Kentucky in 1816 and came to Texas July 4, 1835.[1770] After the Revolution, he received a donation of 640 acres.[1771] He also received one-third league of land from the Bastrop County board.[1772]

VARCINAS, Andreas—Private, Company D, First Regiment Texas Volunteers. He was a native of Texas.[1773] After the Revolution he received one-third league of land issued by Bexar County.[1774]

VERMILLION, Joseph—Private, Company D, First Regiment Texas Volunteers. He arrived at Velasco, January 28, 1836, having been recruited in New Orleans by Captain Turner. He was a member of Captain Hart's Company, January 30, 1836. Most of the men of Captain Hart's Company participated in the Battle of San Jacinto in Captain Richard Roman's Company, but Vermillion fought in Captain Baker's Company. On April 22, 1836 he was one of the captors of Santa Anna. A total of 640 acres of land for having participated in the Battle of San Jacinto were issued May 13, 1873, 320 acres for three months service was issued, and one-third of a league of land was issued. All lands were delivered to Captain Calder.

VINALER, J.—Private, Company B, Volunteers. No information.

VIVEN, John—Private, Company A, First Texas Volunteers. He joined the army before December, 1835.[1775] After the Revolution, he received a donation of 640 acres.[1776] He also received one-third league of land, issued by Harris County.[1777]

VOTAW, Elijah—Sixth Company, Second Regiment Texas Volunteers. He was a native of Arkansas and arrived in Texas in April, 1835. He was a single man.[1778] After the Revolution he received a donation of land.[1779] He also received two-thirds of a league and one labor of land, issued by Montgomery County.[1780]

WADE, John M.—Artillery Corps. He came to Texas before 1833 and received one-fourth league in Austin's Third Colony.[1781] He received one-third league of land from the Harris County board after the Revolution.[1782] He also received a bounty of land.[1783]

WALDRON, E. W.—Company B, Volunteers. He arrived at Velasco January 28, 1836, having been recruited in New Orleans by Captain Turner. He was a member of Captain Hart's Company. Richard Roman succeeded Captain Hart in command and he was in his Company at San Jacinto. He did not apply for the Donation Land Certificate due to him. On December 24, 1858, he made application to the Commissioner of Claims for his donation land. On February 7, 1860 he was issued 640 acres of land for his San Jacinto services. He did not apply for his headright land.

WALKER, James—Sixth Company, Second Regiment Texas Volunteers. He was born in Kentucky in 1790. He came to Texas before July 21, 1824.[1784] He received on July 21, 1824, he received one league in Austin's First Colony.[1785] He received one-third league from the Washington County board after the Revolution.[1786] He also received a bounty.[1787]

WALKER, Martin—Company C, First Regiment Texas Volunteers. He came to Texas in 1835. He received one-third league from the Bastrop County board, after the Revolution.[1788] He also received a bounty of 640 acres.[1789]

WALKER, Phillip—His name is not given in any of the lists, but he was at the Battle because the land files have a certificate for his land, which states that he was. He came to Texas in 1835 and received one-third league and one labor from the San Augustine County board.[1790] He received a donation of land in addition.[1791]

WALKER, W. S.—Company B, Volunteers. He was wounded at the Battle of San Jacinto.

WALLING, Jesse—First Company, Second Regiment Texas Volunteers. He came to Texas in 1834.[1792] He joined the army March 6, 1836.[1793] He received one league and one labor of land from the Nacogdoches County board after the Revolution.[1794] He also received a donation.[1795]

WALNUT, Francis—Fourth Company, Second Regiment Texas Volunteers. He came to Texas before the Declaration of Independence. He received one-third league from the Brazoria County board after the Revolution.[1796] He also received a bounty of land.[1797]

WARDSZISKI, Felix—Company B, Volunteers. He arrived in Texas in January, 1836. He received one-third league of land from the Harris County board after the Revolution.[1798] He also received a bounty of land.[1799]

WARE, William—Captain, Second Regiment Texas Volunteers. He came to Texas in 1831 and received one labor from the Montgomery County board after the Revolution.[1800] He also received a donation of land.[1801]

WARNER, (First name unknown)—Company A, Regulars. No information.

WATERS, George—Company A, First Regiment Texas Volunteers. He came to Texas in December, 1835. He received one league and one labor from the Nacogdoches County board after the Revolution.[1802] He also received a donation.[1803]

WATERS, William—Company F, First Regiment Texas Volunteers. He was born in Virginia in 1787 and came to Texas in April, 1831.[1804] He received one-third league from the Colorado County board after the Revolution.[1805] He also received a donation.[1806]

WATKINS, J. E.—Company D, First Regiment Texas Volunteers. He came to Texas before 1835. He was unmarried and had two slaves.[1807] He received a bounty after the Battle.[1808]

WATSON, Dexter—Eighth Company, Second Regiment Texas Volunteers. He arrived in Texas in 1835. He received one-third league from the San Augustine County board.[1809] He received a bounty of 640 acres in addition.[1810]

WEBB, James—Company A, Regulars. He received a donation after the Battle of San Jacinto.[1811]

WEBB, Thomas H.—Sixth Company, Second Regiment Texas Volunteers. He came to Texas in February, 1835. He received one league of land from the Montgomery County board after the Revolution.[1812] He also received a donation of land.[1813]

WEEDON, George—Company I, Volunteers. He came to Texas in the Fall of 1835. He received one-third league of land from the Washington County board after the Revolution.[1814] He also received a donation of land.[1815]

WELLS, James—Special Scout, Cavalry Corps. He was born in North Carolina in 1817.[1816] He came to Texas before the Declaration of Independence. He received one-third league from the Harris County board after the Revolution. He also received a donation of land.[1817]

WELLS, James—Private, Cavalry Corps. He was born in Missouri and came to Texas September 29, 1833. He was a single man.[1818] After the Revolution he received one-third league and one labor from the Harris County board.[1819] He also received a donation of land.[1820]

WELLS, Lysander—Major, Second Regiment Texas Volunteers. He came to Texas before the Declaration of Independence. He received one-third league of land from the Bexar County board after the Revolution.[1821] He received a donation of land.[1822] He was killed in a duel with Captain William D. Redd at San Antonio in 1840. Both men were killed and the event was deplored through-out the county.[1823]

WELSH, James—Company A, First Regiment Texas Volunteers. He came to Texas before the Declaration of Independence. He received one-third league of land from the Brazoria County board after the Revolution.[1824] He also received a donation of land.[1825]

WEPPLER, Phillip—Company D, First Regiment Texas Volunteers. He came to Texas in 1834. He received one league and one labor of land from the Robertson County board after the Revolution.[1826] He also received a donation of land.[1827]

WESTGATE, Ezra C.—Company A, First Regiment Texas Volunteers. He came to Texas in January, 1836. He received one-third league of land from the Harris County board after the Revolution.[1828] He also received a bounty of land.[1829]

WHARTON, James—Sergeant Company B, Volunteers. He came to Texas in 1836. He received one-third league from the Harris County board after the Revolution.[1830] He also received a bounty of land.[1831]

WHARTON, John A.—Adjutant General. He was born in Virginia. He came to Texas before or during 1830. He was in the Consultation in 1835.[1832] After the Revolution he received one-third league of land from the Brazoria County board.[1833] He also received a donation of land.[1834] He was Secretary of the Navy for a while during Lamar's Administration, but resigned to become a candidate from Brazoria for the first Congress. He was elected, and died during the third session, December 17, 1838.[1835]

WHEELER, S. L.—Sergeant Company B, Volunteers. He came to Texas before the Declaration of Independence. He received one-third league from the Brazoria County board after the Revolution.[1836] He also received a donation of land.[1837]

WHITAKER, Mat G.—First Company, Second Regiment Texas Volunteers. He came to Texas before 1835. He received one-third league from the Nacogdoches County board after the Revolution.[1838] He also received a donation of land.[1839]

WHITE, John C.—Sixth Company, Second Regiment Texas Volunteers. He came to Texas in 1835. He received one-third league from the Liberty County board after the Revolution.[1840] He also received a donation of land.[1841]

WHITE, Joseph—Artillery Corps. He came to Texas in January, 1836. He received one league and one labor from the Nacogdoches County board after the Revolution.[1842] He also received a donation of land.[1843]

WHITE, L. W.—Seventh Company, Second Regiment Texas Volunteers. He came to Texas before the Declaration of Independence. He received a donation of land after the Battle. He also received one-third league of land from the Houston County board.[1844]

WHITEHEAD, N.—Company H, First Regiment Texas Volunteers. (Given only in the Muster Rolls in the Land Office). He was born in Louisiana in 1790 and came to Texas February 2, 1825. He was a bachelor.[1845] He received one-twelfth league of land from the Austin County board after the Revolution.[1846] He also received a donation of land.[1847]

WHITESIDES, E.—Company H, First Regiment Texas Volunteers. He came to Texas in 1825. He received one-third league from the Washington County board after the Revolution.[1848] He also received a donation of land.[1849]

WHITLOCK, Robert—Third Company, Second Regiment Texas Volunteers. He came to Texas in 1827 and he received one-third league of land from the Liberty County board after the Revolution.[1850] He also received a donation of land.[1851]

WILDER, James S.—Company B, Volunteers. He came to Texas in January, 1836.[1852] His heirs received one-third league of land from the Harris County board after the Revolution.[1853]

WILEY, Sam—Sixth Company, Second Regiment Texas Volunteers. No information.

WILKINSON, James—Company A, Regulars. He was born in Hardin County, Kentucky, March 6, 1805 and came to Texas from Tennessee in 1831. He participated in the Battle of San Jacinto, April 21, 1836 as a member of Captain Hill's Company. He enlisted in the army February 2, 1836; he was discharged July 21, 1836. He was issued, on March 21, 1838, one-third of a league of land by the Harrisburg County Board. The certificate was assigned to David P. Coit. On April 30, 1852, 640 acres of land were issued in his name. It had been assigned to James H. Isbell. He re-enlisted in the army and 1280 acres of land were issued in his name for his services. On November 4, 1837 he had sold the rights to the certificate to George Fisher for $100. He died in what is now Burleson County, August 15, 1848 and was buried on his farm.

WILKINSON, Freeman—He arrived in Texas in 1835. He received one league and one labor from the Harris County board, after the Revolution.[1854] He also received a donation of land.[1855]

WILKINSON, John—Second Sergeant, Fifth Company, Second Regiment Texas Volunteers. He was born in 1809 and arrived in Texas January 25, 1836. He had a family of servants.[1856] After the Revolution he received one league and one labor from the Matagorda County board.[1857] He received a donation of land in addition.[1858]

WILKINSON, James G.—Company H, First Regiment Texas Volunteers. He was a native of Tennessee and he came to Texas before 1831.[1859] After the Revolution he received one league and one labor from the Milam County board.[1860] He also received a donation of land.[1861]

WILKINSON, Leroy—Company F, First Regiment Texas Volunteers. He came from Georgia and arrived March 4, 1835. He was a single man.[1862] He received one-third league from the Colorado County board after the Revolution.[1863]

WILLIAMS, Charles—Company C, First Regiment Texas Volunteers. He was a native of Louisiana and he came to Texas in 1834. He was a married man.[1864] In December, 1834, he received one league in Zavala's Colony.[1865] He received one labor from the Jefferson County board after the Revolution.[1866]

WILLIAMS, E. F.—Company B, Volunteers. He came to Texas before the Declaration of Independence. He received one-third league from the Harris County board.[1867]

WILLIAMS, H. R.—Third Company, Second Regiment Texas Volunteers. He was a native of Louisiana. He arrived in Texas in 1834. He was a single man and had three slaves.[1868] After the Revolution he received a donation of land.[1869]

WILLIAMS, W. F.—First Company, Second Regiment Texas Volunteers. He came to Texas before the Declaration of Independence. He received one league of land from the Nacogdoches County board after the Revolution. He also received a donation of land.[1870]

WILLIAMS, R. W.—Company D, First Regiment Texas Volunteers. No information.

WILLIAMSON, J. W.—Cavalry Corps. He was born in Mississippi in 1798 and arrived in Texas in January, 1826. He was a widower and had one son.[1871] He received one-third league of land from the Washington County board after the Revolution.[1872]

WILLOUGHBY, Leaper—Eighth Company, Second Regiment Texas Volunteers. He came to Texas May 2, 1835. He received one-third league from the Brazoria County board after the Revolution.[1873] He also received a bounty of land.[1874]

WILMUTH, Lewis—Eighth Company, Second Regiment Texas Volunteers. He received a bounty of land after the Revolution.[1875]

WILSON, Charles—Cavalry. He came to Texas in 1832. He received one-third league of land from the Matagorda County board after the Revolution.[1876] He also received a bounty of land.[1877]

WILSON, James—Second Company, Second Regiment Texas Volunteers. He was a native of England.[1878] He arrived in Texas in 1835.[1879] He received a bounty of land after the Revolution.[1880] He went on the Mier Expedition in 1842.[1881]

WILSON, Thomas—Company B, Volunteers. He was a native of England and came to Texas before 1835 and was a married man.[1882] October 15, 1835 he received one league in Vehlein's Colony.[1883] He received a bounty of land after the Revolution.[1884]

WILSON, Walker—Company C, First Regiment Texas Volunteers. He was born in Virginia in 1801 and came to Texas March 12, 1835.[1885] May 12, 1835, he received one league in Milam's Colony.[1886] He also received a donation of land.[1887]

WINBURN, McHenry—Company D, First Regiment Texas Volunteers. He was a native of Alabama and came to Texas December 15, 1834.[1888] October 3, 1835, he received one-fourth league of land in Austin's Colony.[1889] He received a donation of land after the Revolution.[1890]

WINN, Walter—Company A, First Regiment Texas Volunteers. He was a native of New England and came to Texas in February, 1835. He was a married man.[1891] He received a bounty of land after the Revolution.[1892]

WINNER, Christian—Company F, First Regiment Texas Volunteers. No information.

WINTERS, J. F.—Second Company, Second Regiment Texas Volunteers. He was born in Tennessee and came to Texas in 1832. He was not married.[1893] After the Revolution he received a donation.[1894]

WINTERS, James W.—Second Company, Second Regiment Texas Volunteers. He was born in Giles County, Tennessee, January 21, 1817. He came to Texas in 1834. He joined the army March 12, 1836.[1895] He received a donation of land after the Revolution.[1896]

WINTERS, William C.—Second Company, Second Regiment Texas Volunteers. He arrived in Texas in 1835.[1897] He joined the army on March 12, 1836.[1898] He received one league in Vehlein's Colony.[1899] After the Battle of San Jacinto he received one labor from the Montgomery County board.[1900] He also received a donation.[1901]

WOOD, Edward B.—Adjutant, Second Regiment Texas Volunteers. He came to Texas after the Declaration of Independence. He received one-third league issued by the Brazoria County board after the Revolution. He also received a donation.[1902]

WOOD, William—Company A, First Regiment Texas Volunteers. He came to Texas before the Declaration of Independence. He came to Texas before the Declaration of Independence. After the Revolution he received one-third league from the Washington County board. He also received a donation.[1903]

WOODS, (First name unknown)—Company B, Volunteers. No information.

WOODWARD, F. M.—Sixth Company, Second Regiment Texas Volunteers. He was born in Tennessee in 1815 and came to Texas before January, 1836. He joined the army on January 14, 1836.[1904] After the Revolution he received a bounty of 320 acres.[1905]

WOOLSEY, Abner W.—Company D, First Regiment Texas Volunteers. He was born in Alabama in 1814 and came to Texas on March 1, 1835. He was not married. He was a farmer.[1906] He received, on October 29, 1835, one-fourth league in Austin's Fifth Colony.[1907] After the Revolution he received a donation.[1908] He also received one-twelfth league from the Colorado County board.[1909]

WRIGHT, Gilbert—Fourth Company, Second Regiment Texas Volunteers. He came to Texas in 1835. After the Revolution he received one-third league from the Matagorda County board.[1910] He also received a donation.[1911]

WRIGHT, Rufus—Company I, Volunteers. He was born in New York and came to Texas on July 11, 1834. He was not married and was a farmer.[1912] After the Revolution he received a donation.[1913]

WYLY, Alfred H.—Captain (His name is given only in the Muster Rolls at the Land Office). He came to Texas before August 30, 1835. He received one-fourth league in Vehlein's Colony.[1914] After the Battle of San Jacinto he received a donation.[1915]

YANCY, John—First Company, Second Regiment Texas Volunteers. He arrived in Texas and held land before August 3, 1835.[1916] He joined the army March 6, 1836.[1917] After the Battle of San Jacinto he received two-thirds of a league and one labor of land from the Nacogdoches County board,[1918] and a donation of 640 acres.[1919]

YARBROUGH, Swanson—First Company, Second Regiment Texas Volunteers. He was a native of Louisiana and arrived in Texas in 1832. He was a single man[1920] and joined the army March 6, 1836.[1921] After the Revolution, he received a donation[1922] and two-thirds of a league and one labor from the Washington County board.[1923]

YOUNG, William F.—Private, Cavalry. He arrived in Texas before February 12, 1835, and received one league of land in Robertson's Colony in Limestone County, February 12, 1835.[1924] After the Revolution, he received a donation,[1925] and one league issued by Montgomery County.[1926]

YORK, Allison—Private, Company D, First Regiment Texas Volunteers. He came to Texas in 1829. After the Revolution, he received one-third league of land issued by Austin County.[1927] He also received a donation of land.[1928]

Part Four
The Sources

1. Quoted by Yoakum, *History of Texas*, II, 373.

2. Van Holst, *Constitutional and Political History of the United States*, II, 570, published in 1888.

3. Van Holst, *Constitutional and Political History of the United States*, II, 570, published in 1888

4. Rives, *The United States and Mexico*, 1821-1849, I, 372.

5. Von Holst, *Constitutional and Political History of the United States*, II, 583.

6. McMaster, *History of the People of the United States*, VI, 266, published in 1907.

7. Wooten (editor), *A Comprehensive History of Texas*, I, X.

8. The application which follows illustrates the more or less stereotyped form of such documents—

   To the Commissioner the Baron de Bastrop Sir— Samuel Isaacs a native of the United Sates of America and now residing in the Province of Texas would respectively represent to you that having emigrated to this country with his family and effects, with the intention of settling himself permanently in the Colony of Stephen F. Austin, established by the authority of the superior Government of the Mexican Nation, he hopes that admitting him with his family as one of the first settlers of the aforesaid Colony you may be pleased to grant him and put him in possession of the quantity of land allowed by law to Colonists, with the understanding that he is ready to settle and cultivate that which may be assigned him, obeying in all cases the existing laws, and to defend the rights, the liberties and the independence of the nation. He therefore prays that you may be pleased to accede to this his petition, wherein he will receive favor and justice.

   Town of San Felipe de Austin, July 9, 1824.
   Samuel Isaacs.
   Translation of Titles in Austin's First Colony, I, 106.

9. Burlage and Hollingsworth, *Abstracts of Land Titles* (1859), page V. "All heads of a family received one League and one Labor, all single men of the age of seventeen years and upwards received one-third of a League. All those who had previously received their League of land as heads of families; and those who received [a] quarter of a League as single persons, received an additional quantity [such] as would make the quantity of land received by them equal to one

league and labor, and one-third or a league."

10. Burlage and Hollingsworth, *Abstracts of Land Titles*, page VI.

11. B & H, *Abstracts*, page VI.

12. This does not include those who were left at Harrisburg to guard the sick and baggage.

13. The month was not given.

14. The month was not given.

15. Book B, List of Applicants for land in Austin's Colonies, 67.

16. Spanish Archives, LIX, 17.

17. B & H, *Abstracts*, 17.

18. Book A, Reports from the Land Boards, A, page 1. (A, page 1 means the first page of the lists of names beginning with A.)

19. Book A, Reports from the Land Boards, A, page 3.

20. B & H, *Abstracts*, 21.

21. Book A, Reports, A, page 3.

22. Thrall, *Pictorial History*, 477.

23. Book A, Reports, A, page 4.

24. Spanish Archives, III, 557.

25. B. & H., *Abstracts*, 21.

26. Book A, Reports, A, page 4.

27. Thrall, *Pictorial History*, 477.

28. Character Certificate issued by Michamps to Allison, May 30, 1835, Spanish Archives.

29. Muster Rolls, 28.

30. Book A, Reports, A, page 1.

31. Book A, Reports, A, page 1.

32. B. & H., *Abstracts*, 21.

33. B. & H., *Abstracts*, 602.

34. B. & H.., *Abstracts* 13, 21 & 17.

35. Book B, Reports, A, page 1.

36. B. & H., *Abstracts*, 17, 21.

37. B. & H., *Abstracts*, 1.

38. B. & H., *Abstracts*, 21.

39. James E. Winston, "Virginia and the Independence of Texas" in *Southwestern Historical Quarterly*, XVI, 279.

40. Book B, Reports, A, page 1.

41. B. & H., *Abstracts*, 17, 21.

42. Book B, Lists of Applicants for Land in Austin's Colonies, 57.

43. Muster Rolls, 208.

44. B. & H., *Abstracts*, 21.

45. Book B, Reports, A, page 2.

46. B. & H., *Abstracts*, 21.

47. B. & H., *Abstracts*, 3.

48. B. & H., *Abstracts*, 19.

49. Book A, Reports, A, page 1.

50. Muster Rolls, 27.

51. Book A, Reports, A, page 1.

52. B. & H., *Abstracts*, 18.

53. Muster Rolls, 115.

54. Muster Rolls, 45.

55. Brown, *History of Texas*, II, 104.

56. Book B, Reports, A, page 1.

57. B. & H., *Abstracts*, 21.

58. Book A, Reports, A, page 3.

59. B. & H., *Abstracts*, 17, 21. 60. Book A, Reports, A, page 1.

61. B. & H., *Abstracts*, 21.

62. Book A, Lists of Applicants for Land in Austin's Colonies, 37.

63. Spanish Archives, IX, 117.

64. Book A, Reports, A, page 3.

65. B. & H., *Abstracts*, 17, 21.

66. Book A, Lists of Applicants for Land in Austin's Colonies, 67.

67. Spanish Archives, IV, 550.

68. Book A, Reports, B, page 11.

69. B. & H., *Abstracts*, 75.

70. Certificate of Character issued by Radford Berry to Howard Bailey, August 17, 1835, Spanish Archives.

71. Muster Rolls, 125.

72. Book A, Reports, B, page 8.

73. Book B, Reports, B, page 3.

74. B. & H., *Abstracts*, 75

75. Book A, Lists of Applicants for Land in Austin's Colonies, 87.

76. Spanish Archives, V, 1410.

77. Book A, Reports, B, page 1.

78. B. & H. *Abstracts*, 67, 77.

79. Brown, *History of Texas*, II, 104.

80. Book B, Lists of Applicants for Land in Austin's Colonies, 9.

81. Spanish Archives, X, 33.

82. Book A, Reports, B, 6.

83. B. & H., *Abstracts*, 69.

84. Brown, *History of Texas*, II, 111.

85. Certificate of Character issued by Wm. Hardin, Judge of Liberty County to Mosely Baker, October 2, 1834, Spanish Archives, Land Office.

86. Book B, Lists of Applicants for Land in Austin's Colonies, 8.

87. Spanish Archives, XXIII, 1161.

88. Thrall, *Pictorial History*, 498.

89. B. & H. *Abstracts*, 77.

90. Thrall, *Pictorial History*, 498.

91. The muster rolls in the Land Office give Capt. Wyly and fifteen men in his Company as participants in the Battle of San Jacinto. The list of names is given, followed by this note:

> I do here certify that owing to an oversight of the proper officers, the within report was not handed to the Adjutant General in time to accompany the report of the Commander-in-chief, and further that Captain Wyly's Company should have been reported as having participated in the Battle of San Jacinto.

"Signed" John A. Wharton

May 18, 1836. Adjt. General.

92. B. & H., *Abstracts*, 73, 77.

93. Book B, Reports, B, page 7.

94. B. & H., *Abstracts*, 68, 76.

95. B. & H., *Abstracts*, 68.

96. B. & H., *Abstracts*, 68.

97. Book A, Reports, B, page 6.

98. B. & H., *Abstracts*, 33.

99. B. & H., *Abstracts*, 69, 76.

100. Barr to Austin, December 5, 1832, Spanish Archives.

101. Brown, *History of Texas*, II, 109.

102. Book B, Reports, B, page 8.

103. B. & H., *Abstracts*, 77.

104. Book A, Reports, B, page 5.

105. Book A, Reports, B, page 7.

106. B. & H., *Abstracts*, 64, 75.

107. B. & H., *Abstracts*, 70.

108. Book A, Reports, B, page 2.

109. B. & H., *Abstracts*, 76.

110. Book A, Reports, B, page 6.

111. B. & H., *Abstracts*, 70.

112. Book A, Reports, B, page 1.

113. B. & H., *Abstracts*, 76.

114. B. & H., *Abstracts*, 23.

115. B. & H., *Abstracts*, 64.

116. Certificate of Character issued by the Alcalde of Nacogdoches to Thomas Y. Beauford, June 2, 1835. Spanish Archives.

117. Spanish Archives, XXXV, 271.

118. Book A, Reports, B, page 9.

119. B. & H., *Abstracts*, 68, 76.

120. B. & H., *Abstracts*, 29.

121. B. & H., *Abstracts*, 36.

122. B. & H., *Abstracts*, 75.

123. Book B, Lists of Applicants for Land in Austin's Colonies, 87.

124. Book A, Reports, B, page 5.

125. Book A, Reports, B, page 4.

126. B. & H., *Abstracts*, 72.

127. Book A, Lists of Applicants for Land in Austin's Colonies, 51.

128. Spanish Archives, XXXIV, 86.

129. B. & H., *Abstracts*, 66.

130. Thrall, *Pictorial History*, 501.

131. B. & H., *Abstracts*, 67.

132. Thrall, *Pictorial History*, 501.

133. Book B, Lists of Applicants for Land in Austin's Colonies, 9.

134. Spanish Archives, XI, 665.

135. Book A, Reports, B, page 1.

136. B. & H., *Abstracts*, 75.

137. B. & H., *Abstracts*, 35.

138. Book A, Reports, B, page 1.

139. B. & H., *Abstracts*, 63, 76.

140. B. & H., *Abstracts*, 71, 72.

141. Thrall, *Pictorial History*, 501.

142. B. & H., *Abstracts*, 68.

143. B. & H., *Abstracts*, 26.

144. B. & H., *Abstracts*, 75.

145. Lists of Applicants for Land in Wavell's Colony 27.

146. B. & H., *Abstracts*, 71, 72.

147. B. & H., *Abstracts*, 33.

148. B. & H., *Abstracts*, 67.

149. B. & H., *Abstracts*, 26.

150. B. & H., *Abstracts*, 67, 75.

151. Book A, Lists of Returns from the Land Boards, B, page 4.

152. B. & H., *Abstracts*, 75.

153. B. & H., *Abstracts*, 35.

154. B. & H., *Abstracts*, 151.

155. Book A, Reports, B, page 5.

156. B. & H., *Abstracts*, 65, 76.

157. Thrall, *Pictorial History*, 501.

158. Book A, Reports, B, page 6.

159. B. & H., *Abstracts*, 66.

160. Book A, Reports, B, page 1.

161. B. & H., *Abstracts*, 76.

162. Book A, Returns from the Land Boards, B, page 2.

163. B. & H., *Abstracts*, 75.

164. Brown, *History of Texas*, II, 104.

165. Book A, Reports, B, page 5.

166. Brown, *History of Texas*, II, 30.

167. Book A, Reports, B, page 5.

168. Certificate of Character issued to George L. Bledsoe by Flores, November 13, 1834.

169. Book A, Reports, B, page 2.

170. B. & H., *Abstracts*, 64.

171. Book A, Reports, B, page 5.

172. B. & H., *Abstracts*, 72.

173. Book A, Reports, B, page 1.

174. B. & H., *Abstracts*, 77.

175. Book A, Reports, B, page 1.

176. B. & H., *Abstracts*, 77.

177. Book B, Lists of Applicants for Land in Austin's Colony, 99.

178. B. & H., *Abstracts*, 70.

179. B. & H., *Abstracts*, 25.

180. B. & H., *Abstracts*, 69, 77.

181. Thrall, *Pictorial History*, 502.

182. Spanish Archives, VI, 1902.

183. Book A, Reports, B, page 8.

184. B. & H., *Abstracts*, 69, 75.

185. Thrall, *Pictorial History*, 502.

186. Book A, Lists of Applicants for Land in Austin's Colonies, 57.

187. Spanish Archives, III, 210.

188. Book A, Reports, B, page 1.

189. B. & H., *Abstracts*, 63

190. Book A, Reports, B, page 6.

191. B. & H., *Abstracts*, 66, 77.

192. Muster Rolls, 28.

193. Book B, Reports, B, page 5.

194. Muster Rolls, 28.

195. Book B, Reports, B, page 5.

196. B. & H., *Abstracts*, 71, 77.

197. Book B, Lists of Applicants for Land in Austin's Colonies, 19.

198. B. & H., *Abstracts*, 64.

199. Book A, Reports, B, page 1.

200. Spanish Archives, XVIII, 247.

201. Muster Rolls, 125.

202. Book A, Reports, B, page 1.

203. B. & H., *Abstracts*, 58, 75.

204. Character Certificate issued by Florey to Box, November 28, 1834, Spanish Archives.

205. Spanish Archives, XXI, 199.

206. Muster Rolls, 125.

207. B. & H., *Abstracts*, 75.

208. Book A, Reports from the Land Boards, B, page 10.

209. Character Certificate issued by Michamps to Box, June 1, 1835. Spanish Archives.

210. Spanish Archives, XX, 234.

211. Muster Rolls, 125.

212. B. & H., *Abstracts*, 77.

213. Book A, Reports, B, page 11.

214. Muster Rolls, 125.

215. Book A, Reports, B, page 11.

216. B. & H., *Abstracts*, 70, 76.

217. Book A, Reports, B, page 6.

218. B. & H., *Abstracts*, 64.

219. Brown, *History of Texas*, II, 104.

220. Book B, Lists of Applicants for Land in Austin's Colonies, 87.

221. B. & H., *Abstracts*, 37.

222. B. & H., *Abstracts*, 67, 76.

223. Book A, Lists of Applicants for Land in Austin's Colonies, 32.

224. Spanish Archives, XI, 573.

225. Book B, Reports, B, page 6.

226. B. & H., *Abstracts*, 77.

227. B. & H., *Abstracts*, 29.

228. Book A, Reports, B, page 1.

229. B. & H., *Abstracts*, 75.

230. Brown, *History of Texas*, II, 104.

231. Book A, Reports, B, page 2.

232. B. & H., *Abstracts*, 76.

233. Lists of Applicants for Land in Austin's Colonies, 95.

234. Spanish Archives, X, 1082.

235. Book A, Reports, B, page 2.

236. B. & H., *Abstracts*, 64, 77.

237. B. & H., *Abstracts*, 68.

238. Certificate of Character issued by the Alcalde of Nacogdoches to Henry Brewer, September 20, 1834.

239. Book B, Reports, B, page 7.

240. B. & H., *Abstracts*, 75.

241. Book A, Reports, B, page 11.

242. B. & H., *Abstracts*, 64, 76.

243. Book B, Reports, B, page 4.

244. Brown, *History of Texas*, II, 30.

245. B. & H., *Abstracts*, 70, 78.

246. Book B, Reports, B, page 4.

247. Book A, Reports, B, page 6.

248. B. & H., *Abstracts*, 66.

249. Book A, Reports, B, page 6.

250. B. & H., *Abstracts*, 68, 75.

251. Thrall, *Pictorial History*, 506.

252. Book A, Reports, B, page 6.

253. B. & H., *Abstracts*, B, page 6.

254. Book A, Reports, B, page 7.

255. B. & H., *Abstracts*, 77.

256. Character Certificate issued by John Bevil to David Brown, September 24, 1834, Spanish Archives.

257. Spanish Archives, XXII, 159.

258. Book A, B, Reports, B, page 12.

259. B. & H., *Abstracts*, 67.

260. Spanish Archives, II, 570.

261. Book A, Reports, B, page 2.

262. Muster Rolls, 28.

263. Book A, Reports, B, page 1.

264. B. & H., *Abstracts*, 66, 76.

265. B. & H., *Abstracts*, 27.

266. B. & H., *Abstracts*, 69, 75.

267. Book A, Reports, B, page 1.

268. B. & H., *Abstracts*, 64, 74.

269. Thrall, *Pictorial History*, 507.

270. Spanish Archives X, 225.

271. Book B, Reports, B, page 9.

272. B. & H., *Abstracts* 72, 75.

273. Book A, Reports, B, page 5.

274. B. & H., *Abstracts*, 75.

275. Lists of Applicants for Land in Wavell's Colony, 30.

276. B. & H., *Abstracts*, 7.

277. Book A, Reports, B, page 3.

278. Book A, Reports, B, page 1.

279. Spanish Archives, XV, 65.

280. B. & H., *Abstracts*, 73, 77.

281. Brown, *History of Texas* II, 104.

282. B. & H., *Abstracts*, 141.

283. B. & H., *Abstracts*, 33.

284. B. & H., *Abstracts*, 69, 75.

285. Book A, Reports.

286. B. & H., *Abstracts*, 175.

287. Thrall, *Pictorial History*, 507.

288. Spanish Archives IV, 682.

289. Thrall, *Pictorial History*, 507.

290. B. & H., *Abstracts*, 65.

291. Book A, Reports, B, page 5.

292. Thrall, *Pictorial History*, 508.

293. B. & H., *Abstracts*, 26.

294. Certificate of Character issued by the Alcalde of Nacogdoches to Burton September 7, 1835, Spanish Archives.

295. Spanish Archives, XIV, 197.

296. Book A, Reports.

297. Brown, *History*, II, 133, 168.

298. Book B, Reports, B, page 1.

299. B. & H., *Abstracts*, 66, 75.

300. B. & H., *Abstracts*, 32.

301. B. & H., *Abstracts*, 77, 168.

302. Book A, Reports, B, page 5.

303. B. & H., *Abstracts*, 63, 75.

304. Certificate of Character issued to Andrew Caddell by the Alcalde of San Augustine, April 3, 1834, Spanish Archives.

305. Spanish Archives, XXII, 405.

306. B. & H., *Abstracts*, 125.

307. Certificate of Character issued to B. F. Cage by Radford Berry, August 11, 1833.

308. Book A, Lists of Applicants for Land in Austin's Colonies, 40.

309. B. & H., *Abstracts*, 117.

310. Thrall, *Pictorial History*, 519.

311. Book A, Lists of Applicants for Land in Austin's Colonies, 17.

312. Spanish Archives, XXXV, 263.

313. Book B, Lists of Applicants for land in Austin's Colonies, 81.

314. B. & H., *Abstracts*, 124.

315. B. & H., *Abstracts*, 124.

316. Book A, Reports, C, page 1.

317. B. & H., *Abstracts*, 126.

318. Certificate of Character issued to Patric Carnal by Radford Berry, Alcalde of Nacogdoches, April 27, 1835.

319. Spanish Archives, XXI, 1095.

320. B. & H., *Abstracts*, 116.

321. Book A, Reports, page 8.

322. Muster Rolls, 115.

323. Book A, Reports, C, page 8.

324. Muster Rolls, 115.

325. B. & H., *Abstracts*, 126.

326. Book A, Reports, C, page 8.

327. Book B, Reports, C, page 2.

328. B. & H., *Abstracts*, 126.

329. Book A, Reports, C, page 7.

330. Book B, Lists of Applicants for Land in Austin's Colony, 99.

331. Book B, Reports, K, page 8.

332. Muster Rolls, 44.

333. Spanish Archives, XXII, 821.

334. Book B, Reports, K, page 5.

335. B. & H., *Abstracts*, 125.

336. Book A, Lists of Applicants for Land in Austin's Colonies, 33.

337. Muster Rolls, 44.

338. B. & H., *Abstracts*, 118.

339. Book A, Reports, C, page 4.

340. B. & H., *Abstracts*, 124.

341. Spanish Archives, XVII, 148.

342. Spanish Archives, XVII, 148.

343. Spanish Archives, XVII, 148.

344. B. & H., *Abstracts*, 119.

345. B. & H., *Abstracts*, 91.

346. B. & H., *Abstracts*, 124.

347. Book A, Reports, A, page 4.

348. B. & H., *Abstracts*, 125.

349. Chaffin to Austin and Williams, April 29, 1834, Spanish Archives.

350. B. & H., *Abstracts*, 125.

351. Book B, Lists of Applicants for Land in Austin's Colony, 87.

352. B. & H., *Abstracts*, 124.

353. Certificate of Character issued to John Chevers by Radford Berry, Alcalde of Nacogdoches, May 25, 1835.

354. B. & H., *Abstracts*, 126.

355. Certificate of Character issued to John F. Chevis by Radford Berry, May 25, 1835, Spanish Archives.

356. B. & H., *Abstracts*, 125.

357. Certificate of Character issued to David Choat by John Stewart, May 6, 1835. Spanish Archives.

358. Spanish Archives, XXV, 1005.

359. B. & H., *Abstracts*, 124.

360. B. & H., *Abstracts*, 117.

361. Certificate of Character issued to Elisha Clapp by Alcalde of Nacogdoches, April 8, 1834.

362. Spanish Archives, XVIII, 95.

363. B. & H., *Abstracts*, 125.

364. T. J. Green, *Mier*, 438.

365. Book A, Lists of Applicants for Land in Austin's Colony, 55.

366. B. & H., *Abstracts*, 125.

367. Certificate of Character issued to John Clarke by Benjamin Lindsey, Alcalde of San Augustine, September 24, 1834. Spanish Archives.

368. B. & H., *Abstracts*, 116.

369. Book A, Reports, C, 7.

370. B. & H., *Abstracts*, 118.

371. B. & H., *Abstracts*, 123.

372. Book A, Reports, C, page 4.

373. B. & H., *Abstracts*, 124.

374. Book B, List of Applicants for Land by Austin's Colonists, 67.

375. B. & H., *Abstracts*, 115.

376. B. & H., *Abstracts*, 83.

377. B. & H., *Abstracts*, 122.

378. Book A, Reports, C, page 1.

379. B. & H., *Abstracts*, 124.

380. Book B, Reports, C, page 7.

381. B. & H., *Abstracts*, 126.

382. Book B, Reports, C, page 7.

383. Book B, Lists of Applicants for Land in Austin's Colonies, 103.

384. Book A, Reports, C, page 1.

385. Book A, Reports, C, page 1.

386. B. & H., *Abstracts*, 125.

387. Certificate of Character issued to James Cole by John Stewart, January 9, 1835. Spanish Archives.   .

388. Book B, List of Applicants for Land in Austin's Colonies, 5.

389. Book A, List of Applicants for Land in Austin's Colonies, 103.

390. Spanish Archives, XIV, 459.

391. B. & H., *Abstracts*, 125.

392. Brown, *History of Texas*, II, 129.

393. Book B, Reports, C, page 6.

394. B. & H., *Abstracts*, 124.

395. Letter written May 18, 1835 by Job S. Collard to Antonio Nixon at Nacogdoches, Spanish Archives XX, 467.

396. Muster Rolls, 44.

397. Spanish Archives, XX, 467.

398. B. & H., *Abstracts*, 125.

399. B. & H., *Abstracts*, 125.

400. Thrall, *Pictorial History*, 528.

401. B. & H., *Abstracts*, 125.

402. Brown, *History of Texas*, II, 103.

403. Thrall, *Pictorial History*, 528.

404. Book A, List of Applicants for Land in Austin's Colonies, 10.

405. B. & H., *Abstracts*, 126.

406. B. & H., *Abstracts*, 83.

407. B. & H., *Abstracts*, 78.

408. B. & H., *Abstracts*, 126.

409. B. & H., *Abstracts*, 83.

410. B. & H., *Abstracts*, 127.

411. Book B, Reports, C, page 2.

412. B. & H., *Abstracts*, 125.

413. Spanish Archives, XVII, 42.

414. B. & H., *Abstracts*, 124.

415. B. & H., *Abstracts*, 125.

416. Book A, Reports, C, page 5.

417. Thrall, *Pictorial History*, 528.

418. B. & H., *Abstracts*, 125.

419. Thrall, *Pictorial History*, 528.

420. B. & H., *Abstracts*, 86.

421. B. & H., *Abstracts*, 125.

422. Wooten (editor), *A Comprehensive History of Texas*, I, 704.

423. B. & H., *Abstracts*, 122.

424. B. & H., *Abstracts*, 87.

425. B. & H., *Abstracts*, 124.

426. B. & H., *Abstracts*, 124.

427. Book A, Reports, C, page 8.

428. B. & H., *Abstracts*, 125.

429. Certificate of Character issued by Joseph Lindley and Elijah Collard to

Lewis Cox on November 16, 1834.

430. Spanish Archives, XX, 751.

431. Book A, Lists of Applicants for Land in Austin's Colonies, 43.

432. Spanish Archives, VI, 89.

433. B. & H., *Abstracts*, 126.

434. T. J. Green, *Mier*, 438.

435. Book A, Reports, C, page 8.

436. B. & H., *Abstracts*, C, page 8.

437. Book A, Reports, C, page 4.

438. B. & H., *Abstracts*, 126.

439. Book A, Reports, C, page 6.

440. B. & H., *Abstracts*

441. Certificate of Character issued by Hotchkiss to Joel B. Crain, May 24, 1835.

442. B. & H., *Abstracts*, 126.

443. Book B, Reports, C, page 1.

444. Spanish Archives, V, 1494.

445. Book B, Reports, C, page 1.

446. B. & H., *Abstracts*, 124.

447. Muster Rolls, 115.

448. Book A, Reports, C, page 4.

449. B. & H., *Abstracts*, 118.

450. Book B, Reports, C, page 4.

451. B. & H., *Abstracts*, 127.

452. T. J. Green, *Mier*, 438.

453. Book B, Reports, C, page 5.

454. T. J. Green, *Mier*, 438.

455. Muster Rolls, 115.

456. Book B, Reports, C, page 1.

457. Muster Rolls, 115.

458. Book B, Reports, C, page 1.

459. Book A, Reports, C, page 5.

460. B. & H., *Abstracts*, 80.

461. B. & H., *Abstracts*, 125.

462. Book A, Reports, C, page 6.

463. Muster Rolls, 133.

464. Book B, Reports, C, page 6.

465. B. & H., *Abstracts*, 122.

466. Lists of Applicants for Land in Milam's Colony, 60.

467. Spanish Archives, XVI, 401.

468. B. & H., *Abstracts*, 125.

469. Spanish Archives II, 360.

470. Book A, Reports, C, page 4.

471. B. & H., *Abstracts*, C, page 4.

472. Lists of Applicants for Land in Austin's Colonies

473. Spanish Archives XVI, 41.

474. B. & H., *Abstracts*, 118.

475. Book A, Reports, C, page 4.

476. B. & H., *Abstracts*

477. Book A, Reports, C, 3.

478. B. & H., *Abstracts*, 124.

479. Book A, Reports, D, page 1.

480. Muster Rolls, 28.

481. Book A, Reports, D, page 1.

482. B. & H., *Abstracts*, 154.

483. Book A, Reports, D, page 3.

484. B. & H., *Abstracts*, 154.

485. Book A, Reports, D, page 4.

486. B. & H., *Abstracts*, 155.

487. Book A, Reports, D, page 3.

488. B. & H., *Abstracts*, 149.

489. Book B, List of Applicants for Land in Austin's Colonies, 13.

490. Spanish Archives, XI, 553.

491. B. & H., *Abstracts*, 150.

492. B. & H., *Abstracts*, 154.

493. Book A, Reports, D, page 3.

494. Muster Rolls, 27.

495. Book A, Reports, D, page 3.

496. B. & H., *Abstracts*, 151.

497. Spanish Archives XIII, 511.

498. Muster Rolls, 133.

499. Book B, Reports, D, page 6.

500. Book A, Reports, D, page 4.

501. Book A, Reports, D, page 4

502. B. & H., *Abstracts*, 154.

503. Book B, Reports, D, page 6.

504. List of Applicants for Land in Wavell's Colony, 89.

505. Spanish Archives, XXII, 421.

506. Book A, Reports, D, page 3.

507. B. & H., *Abstracts*, 154.

508. B. & H., *Abstracts*, 154.

509. Book A, Reports, G, page 4.

510. B. & H., *Abstracts*, 155.

511. Book B, Reports, D, page 2.

512. B. & H., *Abstracts*, 154.

513. Book A, Reports, D, page 3.

514. Book A, Reports, D, page 3.

515. B. & H., *Abstracts*, 154.

516. Thrall, *Pictorial History*, 529.

517. B. & H., *Abstracts*, 155.

518. B. & H., *Abstracts*, 149.

519. B. & H., *Abstracts*, 155.

520. B. & H., *Abstracts*, 127.

521. B. & H., *Abstracts*, 127.

522. B. & H., *Abstracts*, 154.

523. Book B, Lists of Applicants for Land in Austin's Colonies, 101.

524. Book A, Reports, D, page 1.

525. Book A, Reports, D, page 2.

526. B. & H., *Abstracts*, 155.

527. B. & H., *Abstracts*, 131.

528. B. & H., *Abstracts*, 149.

529. Muster Rolls, 44.

530. B. & H., *Abstracts,,* 155

531. Book B, Reports, E, page 6.

532. Lists of Applicants for Land in Austin's Colonies, 85.

533. Book A, Reports, D, page 2.

534. B. & H., *Abstracts*, 152.

535. Book A, Lists of Applicants for Land in Austin's Colonies, 65.

536. Spanish Archives III, 406.

537. Book A, Reports, D, page 3.

538. B. & H., *Abstracts*, 154.

539. Book B, Reports, D, page 5.

540. Book A, Reports, D, page 2.

541. B. & H., *Abstracts*, 154.

542. Book A, Reports, D, page 1.

543. B. & H., *Abstracts*, 154.

544. Book A, Reports, G, page 3.

545. Muster Rolls, 180.

546. Book A, Reports, G, page 3.

547. B. & H., *Abstracts*, 154.

548. T. J. Green, *Mier*, 438

549. Book B, Reports, D, page 1.

550. B. & H., *Abstracts*, 154.

551. T. J. Green, *Mier*, 438.

552. Spanish Archives XX, 403.

553. Book A, Reports, D, page 2.

554. B. & H., *Abstracts*, 154.

555. Book A, Reports, D, page 3.

556. B. & H., *Abstracts*, 154.

557. B. & H., *Abstracts*, 134.

558. B. & H., *Abstracts*, 150, 154.

559. Book B, Reports, E, page 6.

560. Spanish Archives, XIV, 85.

561. Book B, Reports, E, page 6.

562. B. & H., *Abstracts*, 154.

563. Book A, Reports, D, page 1.

564. B. & H., *Abstracts*, 154, 151.

565. Book A, Reports, C, page 1.

566. B. & H., *Abstracts*, 154.

567. Book A, Reports, D, page 3.

568. B. & H., *Abstracts*, 154.

569. Book B, Reports, D, page 3.

570. Certificate of Character, issued by B. Holt to William Earle, August 20, 1835.

571. Book A, Reports, E, page 3.

572. T. J. Green, *Mier*, 438.

573. Book A, Reports, E, page 3.

574. B. & H., *Abstracts*, 166, 168.

575. T. J. Green, *Mier*, 438.

576. List of Applicants for Land in Wavell's Colony, 23.

577. Muster Rolls, 44.

578. Spanish Archives XX, 523.

579. Book A, Reports, E, page 1.

580. B. & H., *Abstracts*, 168, 166.

581. Book A, Reports, E, page 1.

582. B. & H., *Abstracts*, 168.

583. B. & H., *Abstracts*, 156.

584. B. & H., *Abstracts*, 165, 168.

585. B. & H., *Abstracts*, 157.

586. B. & H., *Abstracts*, 168, 166.

587. Muster Rolls, 125.

588. Book B, Reports, E, page 1.

589. B. & H., *Abstracts*, 169.

590. B. & H., *Abstracts*, 157.

591. B. & H., *Abstracts*, 169.

592. List of Applicants for Land in Austin's Colonies, 57.

593. Book B, Reports, E, page 1.

594. B. & H., *Abstracts*, 168, 165.

595. Book A, Reports, E, page 1.

596. B. & H., *Abstracts*, 168.

597. Book A, Reports, E, page 2.

598. B. & H., *Abstracts*, 165.

599. Muster Rolls, 115.

600. Book B, Reports, E, page 2.

601. B. & H., *Abstracts*, 168, 166.

602. Book A, Reports, E, page 2.

603. B. & H., *Abstracts*, 168, 165.

604. Book A, Reports, E, page 1.

605. Spanish Archives, XIV, 623.

606. Book A, Reports, E, page 1.

607. B. & H., *Abstracts*, 168, 166.

608. Book A, Reports, E, page 1.

609. B. & H., *Abstracts*, 166.

610. Book B, Reports, E, page 3.

611. Spanish Archives, X, 17.

612. B. & H., *Abstracts*, 168, 166.

613. Book B, Reports, E, page 3.

614. Muster Rolls, 27.

615. Book B, Reports, E, page 3.

616. B. & H., *Abstracts*, 166.

617. Spanish Archives, XV, 931.

618. B. & H., *Abstracts*, 189.

619. Brown, *History of Texas*, II, 111.

620. Book B, List of Applicants for Land in Austin's Colonies, 89.

621. Book A, Reports, F, page 2.

622. B. & H., *Abstracts*, 188.

623. Book B, List of Applicants for Land in Austin's Colonies, 45.

624. B. & H., *Abstracts*, 189.

625. James E. Winston, "Virginia and the Independence of Texas", in the *Southwestern Historical Quarterly*, XVI, 279.

626. Book A, Reports, F, page 2.

627. B. & H., *Abstracts*, 188.

628. Book A, Reports, F, page 2.

629. B. & H., *Abstracts*, 188.

630. B. & H., *Abstracts*, 184.

631. Muster Roll, 115.

632. Book B, Reports, F, page 2.

633. Muster Rolls, 115.

634. Book B, Reports, F, page 2.

635. Muster Rolls, 115.

636. B. & H., *Abstracts*, 189.

637. Book A, Reports, F, page 3.

638. B. & H., *Abstracts*, 189.

639. Book A, Reports, F, page 3.

640. B. & H., *Abstracts*, 185.

641. Book A, Reports, F, page 1.

642. B. & H., *Abstracts*, 187.

643. Book A, Reports, F, page 2.

644. Spanish Archives, XIV, 292.

645. B. & H., *Abstracts*, 188.

646. T. J. Green, *Mier*, 439.

647. Book A, Reports, F, page 3.

648. Thrall, *Pictorial History*, 539.

649. Book A, Reports, F, page 3.

650. B. & H., *Abstracts*, 187.

651. Thrall, *Pictorial History*, 539.

652. Book A, Lists of Applicants for Land in Austin's Colonies, 82.

653. Book A, Reports, F, page 1.

654. B. & H., *Abstracts*, 188.

655. Muster Rolls, 115.

656. Book B, Reports, F, page 4.

657. Muster Rolls, 115.

658. Book B, Reports, F, page 4.

659. B. & H., *Abstracts*, 189.

660. Book A, Reports, F, page 2.

661. Book A, Reports, F, page 2.

662. Spanish Archives, XL, 1827.

663. B. & H., *Abstracts*, 188.

664. Thrall, *Pictorial History*, 539.

665. B. & H., *Abstracts*, 640.

666. Book B, Reports, F, page 1.

667. B. & H., *Abstracts*, 188.

668. Book B, Reports, F, page 2.

669. B. & H., *Abstracts*, 187.

670. B. & H., *Abstracts*, 184.

671. Book B, Lists of Applicants for Land in Austin's Colonies, 7.

672. B. & H., *Abstracts*, 188.

673. Certificate of Character issued by Radford Berry to John Forbes, April 28, 1835, at Nacogdoches, Spanish Archives.

674. Spanish Archives, XX, 371.

675. Book B, Reports, F, page 3.

676. B. & H., *Abstracts*, 189.

677. Book B, Reports, F, page 5.

678. Muster Rolls, 28.

679. Book B, Reports, F, page 5.

680. B. & H., *Abstracts*, 187.

681. B. & H., *Abstracts*, 187.

682. Book A, Reports, F, page 1.

683. Spanish Archives, VIII, 677.

684. Book A, Reports, F, page 1.

685. Book A, Reports, F, page 5.

686. B. & H., *Abstracts*, 189.

687. Book A, Reports, F, page 1.

688. B. & H., *Abstracts*, 188.

689. Brown, *History of Texas*, II, 111.

690. Thrall, *Pictorial History*, 540.

691. Book B, Reports, F, page 2.

692. B. & H., *Abstracts*, 189.

693. Book B, Lists of Applicants for Land in Austin's Colonies, 93.

694. B. & H., *Abstracts*, 189.

695. B. & H., *Abstracts*, 189.

696. B. & H., *Abstracts*, 173.

697. B. & H., *Abstracts*, 188.

698. Spanish Archives, XIV, 161.

699. B. & H., *Abstracts*, 219.

700. Book A, Reports, G, page 4.

701. B. & H., *Abstracts*, 214.

702. Spanish Archives, III, 40.

703. Book A, Reports, G, page 1.

704. B. & H., *Abstracts*, 212.

705. Book A, List of Applicants for Land in Austin's Colonies, 28.

706. Muster Rolls, 44.

707. Book A, Reports, G, page 2.

708. B. & H., *Abstracts*, 218.

709. B. & H., *Abstracts*, 193, 213 and 218.

710. Book B, Reports, G, page 3.

711. B. & H., *Abstracts*, 219.

712. Brown, *History of Texas*, II, 104.

713. B. & H., *Abstracts*, 194.

714. Book B, Reports, G, page 1.

715. Muster Rolls, 26.

716. Book B, Reports, G, page 1.

717. B. & H., *Abstracts*, 192 and 215.

718. Thomas Gay, to S. F. Austin, May, 1830.

719. Book A, Applicants for Land in Austin's Colonies, 75.

720. Spanish Archives, V, 1498.

721. B. & H., *Abstracts*, 218.

722. Book A, Lists of Applicants for Land in Austin's Colonies, 9.

723. Spanish Archives, III, 194.

724. Book B, Reports, G, page 5.

725. Brown, *History of Texas*, II, 131.

726. Book A, Reports, G, page 3.

727. B. & H., *Abstracts*, 218.

728. List of Applicants for Land in Robinson's Colony, 10.

729. Muster Rolls, 28.

730. Thrall, *Pictorial History*, 541

731. B. & H., *Abstracts*, 318.

732. Certificate of Character, issued to John Gilbert by James Hanks; January, 22, 1835. Spanish Archives.

733. Spanish Archives, XXII, 717.

734. B. & H., *Abstracts*, 219.

735. List of Applicants for Land in Austin's Colonies, 103.

736. Spanish Archives, IV, 782.

737. Book B, Reports, G, page 2.

738. B. & H., *Abstracts*, 212.

739. Book A, Reports, G, page 1.

740. B. & H., *Abstracts*, 218.

741. Book A, Reports, G, page 2.

742. B. & H., *Abstracts*, 218.

743. Brown, *History of Texas*, II, 329.

744. Thrall, *Pictorial History*, 542

745. Muster Rolls, 115.

746. B. & H., *Abstracts*, 219.

747. Book B, Reports, G, page 1.

748. B. & H., *Abstracts*, 218.

749. Book B, Lists of Applicants for Land in Austin's Colonies, 91.

750. Book A, Reports, G, page 2.

751. B. & H., *Abstracts*, 218.

752. Book B, Reports, G, page 6.

753. B. & H., *Abstracts*, 218.

754. John Graham to Austin and Williams, May 27, 1834.

755. Spanish Archives, XXIII, 1507.

756. Book A, Reports, G, page 3.

757. B. & H., *Abstracts*, 214.

758. Book A, Reports, G, page 1.

759. Spanish Archives XVI, 539.

760. Book A, Reports, G, page 1.

761. B. & H., *Abstracts*, 219.

762. Book B, List of Applicants for Land in Austin's Colonies, 95.

763. Book A, Reports, G, page 3.

764. B. & H., *Abstracts*, 216.

765. Book B, List of Applicants for Land in Austin's Colonies, 73.

766. Spanish Archives X, 141.

767. B. & H., *Abstracts*, 218.

768. Book A, Reports, G, page 5.

769. Spanish Archives, X, 333.

770. Book A, Reports, G, page 5.

771. Book B, List of Applicants for Land in Austin's Colonies, 33.

772. B. & H., *Abstracts*, 218.

773. Book B, List of Applicants for Land in Austin's Colonies, 23.

774. Book A, Reports, G, page 2.

775. B. & H., *Abstracts*, 218.

776. Muster Rolls, 115.

777. Book A, Reports, G, page 2.

778. Muster Rolls, 115.

779. Book A, Reports, G, page 2.

780. B. & H., *Abstracts*, 218.

781. Brown, *History of Texas*, II, 104.

782. Certificate of Character issued by Radford Berry to A. Greenlaw, September 25, 1835, Spanish Archives.

783. Muster Rolls, 28.

784. B. & H., *Abstracts*, 216.

785. Book A, Reports, G, page 5.

786. B. & H., *Abstracts*, 219.

787. B. & H., *Abstracts*, 192, 215, and 218.

788. List of Applicants for Land in Robertson's Colony, 6.

789. Muster Rolls, 28.

790. B. & H., *Abstracts*, 212.

791. Book A, Reports, G, page 1.

792. B. & H., *Abstracts*, 218.

793. Book A, Reports, F, page 2.

794. B. & H., *Abstracts*, 218.

795. Book A, Reports, I, page 6.

796. B. & H., *Abstracts*, 262.

797. Book A, List of Applicants for Land in Austin's Colonies, 34.

798. Spanish Archives, IX, 205.

799. B. & H., *Abstracts*, 256.

800. List of Applicants for Land in Austin's Colonies, 91.

801. Brown, *History of Texas*, II, 30.

802. B. & H., *Abstracts*, 268.

803. B. & H., *Abstracts*, 246.

804. Book A, List of Applicants for Land in Austin's Colonies, 77.

805. Spanish Archives, XIV, 535.

806. Book A, Reports, H.

807. B. & H., *Abstracts*, 258.

808. Spanish Archives, IV, 1046.

809. B. & H., *Abstracts*, 266.

810. Book B, Reports, H, page 3.

811. B. & H., *Abstracts*, 268.

812. Book A, Reports, H, page 11.

813. B. & H., *Abstracts*, 267.

814. Book A, Reports, H, page 8.

815. B. & H., *Abstracts*, 266.

816. Book A, Reports, H, page 6.

817. Book A, Reports, M, page 1.

818. B. & H., *Abstracts*, 266.

819. Book B, Lists of Applicants for Land in Austin's Colonies, 7.

820. Book A, Reports, M, page 3.

821. B. & H., *Abstracts*, 267.

822. Brown, *History of Texas*, II, 182.

823. Book A, Reports, H, page 1.

824. Spanish Archives, XXXIX, 1024.

825. Book A, Reports, H, page 1.

826. B. & H., *Abstracts*, 266.

827. Muster Rolls, 115.

828. B. & H., *Abstracts*, 267.

829. Certificate of Character issued by A. S. Hotchkiss to John Harmon, June 13, 1835. Spanish Archives.

830. Spanish Archives, XIV, 81.

831. B. & H., *Abstracts*, 266.

832. Certificate of Character, issued by Wm. Woods to Benjamin J.

Harper, December 19, 1834. Spanish Archives.

833. Spanish Archives, XX, 155.

834. B. & H., *Abstracts*, 268.

835. Book B, Reports, H, page 6.

836. B. & H., *Abstracts*, 269.

837. Book A, List of Applicants for Land in Austin's Colonies, 89.

838. Certificate of Character, issued by Nathan Davis to James Harris, October 5, 1835, Spanish Archives.

839. Spanish Archives, XXIV, 211.

840. B. & H., *Abstracts*, 266.

841. List of Applicants for Land in Robinson's Colony, 26.

842. Book B, Reports, H, page 3.

843. Book A, List of Applicants for Land in Austin's Colonies, 79.

844. Spanish Archives, VII, 56.

845. B. & H., *Abstracts*, 258.

846. Certificate of Character, G. Thompson to A. L. Harrison, September 1, 1835. Spanish Archives.

847. B. & H., *Abstracts*, 266.

848. Book B, List of Applicants for Land in Austin's Colonies, 75.

849. B. & H., *Abstracts*, 265.

850. Book B, List of Applicants for Land in Robinson's Colony, 26.

851. Book A, Reports, H, page 7.

852. B. & H., *Abstracts*, 266.

853. B. & H., *Abstracts*, 220.

854. Book A, Reports, H, page 1.

855. B. & H., *Abstracts*, 266.

856. List of Applicants for Land in Austin's Colonies, 32.

857. Spanish Archives, VI, 1854.

858. B. & H., *Abstracts*, 267.

859. List of Applicants for Land in Austin's Colonies, 32.

860. B. & H., *Abstracts*, 266.

861. List of Applicants for Land in Austin's Colonies, 89.

862. B. & H., *Abstracts*, 267.

863. Book A, Reports, H, page 2.

864. B. & H., *Abstracts*, 266.

865. Thrall, *Pictorial History*, 550.

866. Spanish Archives, VII, 242.

867. Thrall, *Pictorial History*, 551.

868. Book A, Reports, H, page 1.

869. B. & H., *Abstracts*, 267.

870. Book A, List of Applicants for Land in Austin's Colonies, 107.

871. B. & H., *Abstracts*, 267.

872. Book B, Reports, H, page 4.

873. B. & H., *Abstracts*, 267.

874. B. & H., *Abstracts*, 259.

875. Book A, Reports, M, page 6.

876. Spanish Archives, XV, 883.

877. Book A, Reports, M, page 6.

878. B. & H., *Abstracts*, 266.

879. Book A, Reports, H, page 5.

880. B. & H., *Abstracts*, 266.

881. B. & H., *Abstracts*, 265.

882. B. & H., *Abstracts*, 260.

883. Book B, List of Applicants for

Land in Austin's Colonies, 47.

884. Book B, Reports, M, page 3.

885. B. & H., *Abstracts*, 266.

886. Book A, List of Applicants for Land in Austin's Colonies, 7.

887. Book B, Reports, H, page 6.

888. Spanish Archives, IV, 658.

889. Book A, Reports, H, page 6.

890. B. & H., *Abstracts*, 268.

891. B. & H., *Abstracts*, 266.

892. Spanish Archives, XVI, 337.

893. B. & H., *Abstracts*, 266.

894. Book A, Reports, H, page 5.

895. B. & H., *Abstracts*, 266.

896. Book A, Reports, H, page 6.

897. Book B, Reports, H, page 4.

898. B. & H., *Abstracts*, 266.

899. List for Applicants for Land in Robertson's Colony, 15.

900. Book A, Reports, H, page 6.

901. B. & H., *Abstracts*, 266.

902. Thrall, *Pictorial History*, 555

903. Spanish Archives, XV, 495.

904. B. & H., *Abstracts*, 256.

905. Book A, Reports, H, page 6.

906. B. & H., *Abstracts*, 267.

907. List of Applicants for Land in Austin's Colonies, 89.

908. Spanish Archives, V, 1590.

909. B. & H., *Abstracts*, 266.

910. Book B, Reports, B, page 16.

911. Book B, Reports, B, page 17.

912. B. & H., *Abstracts*, 267.

913. B. & H., *Abstracts*, 222 and 266.

914. B. & H., *Abstracts*, 256.

915. Book A, List of Applicants for Land in Austin's Colonies, 53.

916. Spanish Archives, III, 146.

917. B. & H., *Abstracts*, 267.

918. B. & H., *Abstracts*, 263 and 266.

919. List of Applicants for Land in Wavell's Colony, 107.

920. Book A, Reports, M, page 5.

921. Certificate of Character, issued to Alexander Horton, September 24, 1834, Spanish Archives.

922. Spanish Archives, XXII, 295.

923. Book A, Reports, I, page 8.

924. B. & H., *Abstracts*, 265.

925. Book B, Reports, H, page 7.

926. B. & H., *Abstracts*, 266.

927. Thrall, *Pictorial History*, 555-568.

928. Book A, Reports, H, page 1.

929. Muster Rolls, 28.

930. Book A, Reports, H, page 1.

931. Book A, List of Applicants for Land in Austin's Colony, I, page 7.

932. Spanish Archives, XIII, 697.

933. B. & H., *Abstracts*, 266.

934. Book A, List of Applicants for Land in Austin's Colonies, 101.

935. Book B, Reports, H, page 1.

936. B. & H., *Abstracts*, 266.

937. Book B, List of Applicants for Land in Austin's Colonies, 153.

938. Book A, Reports, H, page 4.

939. B. & H., *Abstracts*, 266.

940. List of Applicants for Land in Austin's Colonies, 47.

941. Book B, Reports, J, page 1.

942. B. & H., *Abstracts*, 271, 272.

943. Book A, Reports, J, page 2.

944. B. & H., *Abstracts*, 271, 272.

945. "Extracts from a Biographical sketch of Captain John Ingraham", in the *Quarterly of the Texas State Historical Association*, VI, 320.

946. Book A, Reports, J, page 2.

947. B. & H., *Abstracts*, 271, 272.

948. "Extracts from a Biographical sketch of Captain John Ingraham", in the *Quarterly of the Texas State Historical Association*, VI, 320.

949. Certificate of Character, issued to James T. P. Irvine by Mr. B. Holt, Spanish Archives.

950. B. & H., *Abstracts*, 272.

951. Book A, Reports, J, page 4.

952. B. & H., *Abstracts*, 271, 272.

953. Book A, Reports, J, page 1.

954. B. & H., *Abstracts*, 271, 272.

955. Book A, Reports, J, page 1.

956. B. & H., *Abstracts*, 271, 272.

957. Thrall, *Pictorial History*, 570.

958. Book B, Lists of Applicants for Land in Austin's Colonies, 77.

959. Thrall, *Pictorial History*, 570.

960. Book B, Reports, J, page 2.

961. B. & H., *Abstracts*, 288.

962. Brown, *History of Texas*, II, 168.

963. Thrall, *Pictorial History*, 570.

964. Spanish Archives, XVI, 193.

965. B. & H., *Abstracts*, 286, 289.

966. B. & H., *Abstracts*, 286.

967. B. & H., *Abstracts*, 286.

968. Book B, List of Applicants for Land in Austin's Colonies, 85.

969. B. & H., *Abstracts*, 289.

970. Book B, Reports, J, page 1.

971. B. & H., *Abstracts*, 284.

972. Certificate of Character, issued to J. M. Jett by W. A. Pattillo, December 21, 1834.

973. Spanish Archives, XXII, 231.

974. B. & H., *Abstracts*, 288.

975. List of Applicants for Land in Robinson's Colony, 34.

976. Certificate of Character issued to Steven Jett by W. A. Patillo, December 20, 1834.

977. Spanish Archives, XXII, 231.

978. Book B, Reports, J, page 1.

979. B. & H., *Abstracts*, 277, 288.

980. Book B, Reports, J, page 1.

981. B. & H., *Abstracts*, 286, 288.

982. Book A, Reports, J, page 3.

983. B. & H., *Abstracts*, 289.

984. B. & H., *Abstracts*, 274, 288.

985. Certificate of Character issued to James R. Johnson by the Alcalde, September 26, 1834, Spanish Archives.

986. Book B, Reports, J, page 1.

987. B. & H., *Abstracts*, 289, 285.

988. Book B, Reports, J, page 1.

989. B. & H., *Abstracts*, 289, 284.

990. Book A, Reports, J, page 2.

991. B. & H., *Abstracts*, 288.

992. Thrall, *Pictorial History*, 575.

993. Book B, Reports, J, page 2.

994. B. & H., *Abstracts*, 204.

995. Thrall, *Pictorial History*, 576.

996. B. & H., *Abstracts*, 284, 288.

997. Certificate of Character issued to Geo. W. Jones by the Alcalde, September 23, 1834.

998. Book A, Reports, J, page 4.

999. B. & H., *Abstracts*, 284, 289.

1000. Book A, Reports, J, page 3.

1001. Book A, Reports, K, page 2.

1002. B. & H., *Abstracts*, 302.

1003. Thrall, *Pictorial History*, 578

1004. Book A, Reports, K, page 2.

1005. B. & H., *Abstracts*, 302.

1006. Thrall, *Pictorial History*, 578

1007. Book B, List of Applicants for Land in Austin's Colonies, 57.

1008. Book B, Reports, K, page 1.

1009. B. & H., *Abstracts*, 303.

1010. Spanish Archives, XIII, 693.

1011. Book A, Reports, K, page 1.

1012. B. & H., *Abstracts*, 302.

1013. Book B, Reports, K, page 1.

1014. B. & H., *Abstracts*, 298.

1015. Book A, Reports, K, page 2.

1016. B. & H., *Abstracts*, 302.

1017. B. & H., *Abstracts*, 300.

1018. Certificate of Character issued to William Kimbrough by A. Hotchkiss, primary judge of San Augustine, February, 1831.

1019. Book B, Reports, K, page 3.

1020. B. & H., *Abstracts*, 302.

1021. Book A, Reports, K, page 2.

1022. B. & H., *Abstracts*, 303

1023. Certificate of Character issued to William King by A. Hotchkiss, primary judge of San Augustine, August 18, 1834.

1024. Book B, Reports, K, page 1.

1025. B. & H., *Abstracts*, 302.

1026. Book B, List of Applicants for Land in Austin's Colonies, 69.

1027. Book A, Reports, K, page 3.

1028. B. & H., *Abstracts*, 303.

1029. Book A, Reports, K, page 1.

1030. B. & H., *Abstracts*, 301.

1031. Book B, Reports, K, page 2.

1032. B. & H., *Abstracts*, 303.

1033. Book A, Reports, K, page 3.

1034. B. & H., *Abstracts*, 303.

1035. Book A, Reports, L, page 1.

1036. B. & H., *Abstracts*, 329.

1037. B. & H., *Abstracts*, 310 and 324.

1038. Thrall, *Pictorial History*, 581.

1039. B. & H., *Abstracts*, 326.

1040. Thrall, *Pictorial History*, 581-2.

1041. B. & H., *Abstracts*, 325.

1042. Certificate of Character, issued to George Lamb by Antonio Nixon, September 15, 1835, Spanish Archives.

1043. Muster Rolls, 46.

1044. Book A, Reports, L, page 5.

1045. B. & H., *Abstracts*, 329.

1046. Book B, Reports, L, page 4.

1047. B. & H., *Abstracts*, 329.

1048. C. W. Raines, *Bibliography*, 136.

1049. Book B, Reports, L, page 6.

1050. B. & H., *Abstracts*, L, page 6.

1051. C. W. Raines, *Bibliography of Texas*, 136.

1052. Muster Rolls, 44.

1053. Book B, Reports, L, page 3.

1054. B. & H., *Abstracts*, 640.

1055. B. & H., *Abstracts*, 329.

1056. Book A, Reports, L, page 5.

1057. Muster Rolls, 44.

1058. Book A, Reports, L, page 5.

1059. B. & H., *Abstracts*, 329.

1060. Book A, List of Applicants for Land in Austin's Colonies.

1061. Spanish Archives, III, 498.

1062. Book A, Reports, L, page 2.

1063. B. & H., *Abstracts*, 324, 329.

1064. Book B, Reports, L, page 5.

1065. B. & H., *Abstracts*, 325.

1066. Certificate of Character, issued to E. O. Legrand by the Alcalde of San Augustine, December 3, 1834, Spanish Archives.

1067. Spanish Archives XXII, 463.

1068. Book A, Reports, L, page 5.

1069. B. & H., *Abstracts*, 330.

1070. Book A, Reports, L, page 2.

1071. B. & H., *Abstracts*, 329.

1072. Book B, Reports, L, page 2.

1073. B. & H., *Abstracts*, 323.

1074. Book A, Reports, L, page 2.

1075. B. & H., *Abstracts*, 329.

1076. Brown, *History*, II, 103, 168.

1077. B. & H., *Abstracts*, 329.

1078. Book B, Reports, L, page 6.

1079. B. & H., *Abstracts*, 330, 325.

1080. B. & H., *Abstracts*, 323.

1081. Book B, List of Applicants for Land in Austin's Colonies, 31.

1082. Book A, Reports, L, page 5.

1083. B. & H., *Abstracts*, 330.

1084. B. & H., *Abstracts*, 305, 330, 325.

1085. Book A, List of Applicants for Land in Austin's Colonies, 106.

1086. B. & H., *Abstracts*, 306.

1087. Spanish Archives, IX, 273.

1088. B. & H., *Abstracts*, 306, 323.

1089. Spanish Archives, XXXIV, 16.

1090. B. & H., *Abstracts*, 330, 327.

1091. Book A, Reports, L, page 1.

1092. Spanish Archives, IV, 642.

1093. Book A, Reports, L, page 1.

1094. B. & H., *Abstracts*, 329.

1095. Book B, Reports, L, page 1.

1096. Book A, Reports, L, page 4.

1097. B. & H., *Abstracts*, 323.

1098. Brown, *History of Texas*, 142.

1099. Certificate of Character, issued to Benjamin Lindsay by E. Raines, September 22, 1832, Spanish Archives.

1100. Spanish Archives, XXII, 279.

1101. Book B, Reports, L, page 4.

1102. B. & H., *Abstracts*, 330.

1103. Certificate of Character, issued to William M. Logan by William Harding, April 17, 1835.

1104. Spanish Archives, XXI, 1425.

1105. B. & H., *Abstracts*, 330.

1106. B. & H., *Abstracts*, 309, 330, 328.

1107. Book A, Reports, L, page 6.

1108. B. & H., *Abstracts*, 29.

1109. Book A, Reports, L, page 6.

1110. Book A, Reports, L, page 2.

1111. B. & H., *Abstracts*, 329.

1112. Book A, List of Applicants for Land in Austin's Colonies, 57.

1113. Book A, Reports, L, page 4.

1114. B. & H., *Abstracts*, 329.

1115. Book A, List of Applicants for Land in Austin's Colonies, 87.

1116. Spanish Archives, V, 1542.

1117. Book A, Reports, M, page 3.

1118. B. & H., *Abstracts*, 386.

1119. Book B, Reports, M, page 8.

1120. B. & H., *Abstracts*, 386.

1121. Book A, List of Applicants for Land in Austin's Colonies, 79.

1122. Spanish Archives, VI, 2062.

1123. Book A, Reports, M, page 6.

1124. B. & H., *Abstracts*, 386.

1125. B. & H., *Abstracts*, 335.

1126. Book A, List of Applicants for Land in Austin's Colonies, 42.

1127. Spanish Archives, I, 58.

1128. B. & H., *Abstracts*, 379.

1129. Book A, List of Applicants for Land in Austin's Colonies, 71.

1130. Book B, Reports, M, page 6.

1131. B. & H., *Abstracts*, 388.

1132. Book A, List of Applicants for Land in Austin's Colonies, 107.

1133. Book A, Reports, M, page 8.

1134. B. & H., *Abstracts*, 386.

1135. B. & H., *Abstracts*, 340, 388.

1136. Book A, Reports, M, page 2.

1137. T. J. Green, *Mier*, 440.

1138. Book B, Reports, M

1139. B. & H., *Abstracts*, 386.

1140. T. J. Green, *Mier*, 440.1141. Certificate of Character, issued to Thomas Maxwell by R. O. Mc-Daniel, March 28, 1835. Spanish Archives.

1142. Book A, Reports, M, page 5.

1143. B. & H., *Abstracts*, 382.

1144. B. & H., *Abstracts*, 374.

1145. Book A, Reports, M, page 6.

1146. B. & H., *Abstracts*, 388.

1147. Book A, List of Applicants for Land in Austin's Colonies, 89.

1148. Spanish Archives, IV, 618.

1149. B. & H., *Abstracts*, 388.

1150. Book B, List of Applicants for Land in Austin's Colonies, 103.

1151. Book A, Reports, M, page 2.

1152. B. & H., *Abstracts*, 380.

1153. Folson, *Mexico in 1842*, 249.

1154. Book B, List of Applicants for Land in Austin's Colonies, 73.

1155. Book A, Reports, B, page 8.

1156. B. & H., *Abstracts*, 386.

1157. T. J. Green, *Mier*, 441.

1158. Book A, Reports, B

1159. T. J. Green, *Mier*, 441.

1160. B. & H., *Abstracts*, 345, 386.

1161. B. & H., *Abstracts*, 335.

1162. Book A, List of Applicants for Land in Austin's Colonies

1163. Spanish Archives, VI, 1658.

1164. B. & H., *Abstracts*, 386.

1165. Book A, List of Applicants for Land in Austin's Colonies, 3.

1166. Book B, Reports, M, page 5.

1167. Spanish Archives, XII, 427.

1168. Book B, Reports, M, page 3.

1169. B. & H., *Abstracts*, 387.

1170. Book B, List of Applicants for Land in Austin's Colonies.

1171. B. & H., *Abstracts*, 388.

1172. Thrall, *Pictorial History*, 588-99.

1173. Book A, Reports, M, page 2.

1174. B. & H., *Abstracts*, M, page 2.

1175. Book B, Reports, M, page 10.

1176. B. & H., *Abstracts*, 386.

1177. B. & H., *Abstracts*, 339, 640.

1178. Book A, Reports, M, page 11.

1179. B. & H., *Abstracts*, 388.

1180. B. & H., *Abstracts*, 382.

1181. Certificate of Character, issued to Thomas McIntire by the Alcalde of San Augustine, September 1, 1835, Spanish Archives.

1182. Book A, Reports, M, page 5.

1183. B. & H., *Abstracts*, 388.

1184. Book B, Lists of Applicants for Land in Austin's Colonies, 5.

1185. Book A, Reports, M, page 5.

1186. B. & H., *Abstracts*, 386.

1187. Book B, Reports, M, page 9.

1188. B. & H., *Abstracts*, 388.

1189. Book A, Reports, M, page 9.

1190. B. & H., *Abstracts*, 386.

1191. B. & H., *Abstracts*, 387.

1192. Book B, Reports, M, page 9.

1193. Book A, Reports, M, page 7.

1194. B. & H., *Abstracts*, 388.

1195. Spanish Archives, II, 342.

1196. Book A, Reports, M, page 1.

1197. B. & H., *Abstracts*, 388.

1198. Book B, List of Applicants for Land in Austin's Colonies, 95.

1199. Book B, Reports, B, page 7.

1200. B. & H., *Abstracts*, 388.

1201. Book A, Reports, M, page 8.

1202. B. & H., *Abstracts*, 382.

1203. Book B, List of Applicants for Land in Austin's Colonies, 45.

1204. Book A, Reports, M, page 5.

1205. B. & H., *Abstracts*, 386.

1206. Brown, *History*, II, 141, 215.

1207. Book A, List of Applicants for Land in Austin's Colonies, 59.

1208. Spanish Archives, VII, 25.

1209. Book B, Reports, M, page 3.

1210. B. & H., *Abstracts*, 386.

1211. Book B, List of Applicants for Land in Austin's Colonies, 79.

1212. Book B, Reports, M, page 3.

1213. B. & H., *Abstracts*, 386.

1214. Book B, List of Applicants for Land in Austin's Colonies, 79.

1215. Spanish Archives, XXI, 1183.

1216. Book A, Reports, M, page 3.

1217. B. & H., *Abstracts*, 387.

1218. B. & H., *Abstracts*, 387.

1219. B. & H., *Abstracts*, 388.

1220. Book A, List of Applicants for Land in Austin's Colonies, 89.

1221. Muster Rolls, 28.

1222. B. & H., *Abstracts*, 386.

1223. Book B, Reports, M, page 4.

1224. Book B, Reports, M, page 5.

1225. B. & H., *Abstracts*, 388.

1226. Thrall, *Pictorial History*, 592.

1227. Book B, Reports, M, page 3.

1228. B. & H., *Abstracts*, 386.

1229. B. & H., *Abstracts*, 386.

1230. Book B, Reports, O, page 3.

1231. Book A, List of Applicants for Land in Austin's Colonies, 55.

1232. Spanish Archives, IV, 910.

1233. B. & H., *Abstracts*, 378.

1234. B. & H., *Abstracts*, 332, 380.

1235. Book A, List of Applicants for Land in Austin's Colonies, 25.

1236. B. & H., *Abstracts*, 386.

1237. B. & H., *Abstracts*, 339, 375.

1238. Book A, List of Applicants for Land in Austin's Colonies, 34.

1239. Spanish Archives, VI, 1846.

1240. B. & H., *Abstracts*, 386.

1241. Book A, List of Applicants for Land in Austin's Colonies, 23.

1242. Book A, Reports, M, page 12.

1243. Book B, Reports, O, page 6.

1244. Spanish Archives, XVI, 225.

1245. Book B, Reports, O, page 6.

1246. B. & H., *Abstracts*, 386.

1247. B. & H., *Abstracts*, 388.

1248. Spanish Archives, I, 28.

1249. B. & H., *Abstracts*, 375.

1250. Book A, Reports, M, page 1.

1251. Spanish Archives, XVIII, 227.

1252. Book A, Reports, M, page 1.

1253. B. & H., *Abstracts*, 386.

1254. Book A, Reports, M, page 5.

1255. B. & H., *Abstracts*, 386.

1256. Book B, List of Applicants for Land in Austin's Colonies, 63.

1257. Book A, Reports, M, page 8.

1258. B. & H., *Abstracts*, 386.

1259. Book A, List of Applicants for Land in Austin's Colonies, 87.

1260. Spanish Archives, IV, 526.

1261. B. & H., *Abstracts*, 386.

1262. Book A, List of Applicants for

Land in Austin's Colonies, 83.

1263. Spanish Archives, XV, 827.

1264. B. & H., *Abstracts*, 386.

1265. J. W. Winston, "Virginia and the Independence of Texas", *Southwestern Historical Quarterly*, XVI, 279.

1266. Spanish Archives, XV, 763.

1267. Book B, Reports, O, page 4.

1268. B. & H., *Abstracts*, 387.

1269. Muster Rolls, 133.

1270. B. & H., *Abstracts*, 386.

1271. Book B, Reports, M, page 1.

1272. Book A, List of Applicants for Land in Austin's Colonies, 93.

1273. Spanish Archives, VIII, 50.

1274. B. & H., *Abstracts*, 386.

1275. Book B, Reports, M, page 1.

1276. B. & H., *Abstracts*, 388.

1277. Book A, List of Applicants for Land in Austin's Colonies, 3.

1278. Spanish Archives, XXXIV.

1279. B. & H., *Abstracts*, 386.

1280. B. & H., *Abstracts*, 640.

1281. Certificate of Character, issued by Wm. Hardin to Isaac Moreland, December 26, 1834. Spanish Archives.

1282. Spanish Archives, XX, p. 551.

1283. Book B, Reports, S, 15.

1284. B. & H., *Abstracts*, 382.

1285. Book B, Reports, M, page 9.

1286. B. & H., *Abstracts*, page 387.

1287. List of Applicants for land in Wavell's Colony, 34.

1288. B. & H., *Abstracts*, 780.

1289. Book B, List of Applicants for Land in Austin's Colonies, 87.

1290. B. & H., *Abstracts*, 386.

1291. Book B, Reports, O, page 4.

1292. Book A, List of Applicants for Land in Austin's Colonies, 67.

1293. Muster Rolls, 44.

1294. Book B, Reports, M, page 3.

1295. B. & H., *Abstracts*, 388.

1296. J. W. Winston, "Kentucky and the Independence of Texas", in the *Southwestern Historical Quarterly*, XVI, 33.

1297. Book A, Reports, M, page 8.

1298. Brown, *History*, II, page 30.

1299. B. & H., *Abstracts*, 386.

1300. Book A, Reports, M, page 6.

1301. B. & H., *Abstracts*, 386.

1302. Spanish Archives, XIX, 693.

1303. Book A, Reports, N, page 1.

1304. B. & H., *Abstracts*, 378.

1305. Book A, Reports, N, page 2.

1306. Spanish Archives, XXII, 789.

1307. Book A, Reports, N, page 2.

1308. B. & H., *Abstracts*, 398.

1309. Book A, List of Applicants for Land in Austin's Colonies, 101.

1310. Spanish Archives, IX, 169.

1311. B. & H., *Abstracts*, 398.

1312. Book A, Reports, N, page 1.

1313. B. & H., *Abstracts*, 398.

1314. B. & H., *Abstracts*, 396.

1315. Book B, Reports, N, page 2.

1316. Spanish Archives, II, 328.

1317. Book B, Reports, N, page 2.

1318. B. & H., *Abstracts*, 398.

1319. Book B, List of Applicants for Land in Austin's Colonies, 76.

1320. Book A, Reports, N, page 1.

1321. B. & H., *Abstracts*, 398.

1322. Book A, Reports, N, page 2.

1323. B. & H., *Abstracts*, 399.

1324. Book A, Reports, N, page 2.

1325. Book B, Reports, D, page 1.

1326. B. & H., *Abstracts*, page 1.

1327. Book A, Reports, O, page 1.

1328. B. & H., *Abstracts*, 406.

1329. Book B, Reports, O, page 1.

1330. B. & H., *Abstracts*, 406.

1331. Muster Rolls, 131.

1332. B. & H., *Abstracts*, 406.

1333. Book A, Reports, O, page 1.

1334. Spanish Archives, XLIV, 114.

1335. Book A, Reports, O, page 1.

1336. B. & H., *Abstracts*, 405.

1337. Book A, Lists of Applicants for Land in Austin's Colonies, 3.

1338. Spanish Archives, IX, 137.

1339. B. & H., *Abstracts*, 405.

1340. Book A, Reports, O, page 1.

1341. B. & H., *Abstracts*, 405.

1342. Book B, Reports, O, page 1.

1343. B. & H., *Abstracts*, O, page 1.

1344. Book A, Reports

1345. B. & H., *Abstracts*, 434.

1346. Book A, Reports, P, page 4.

1347. B. & H., *Abstracts*, 434, 431.

1348. List of Applicants for Land in Austin's Colonies, 59.

1349. Certificate of Character issued by Wm. Harlen to W. Pace, December 27, 1834.

1350. Spanish Archives, XX, 375.

1351. B. & H., *Abstracts*, 433.

1352. Book B, Reports, P, page 2.

1353. Spanish Archives, XX, 375.

1354. Book B, Reports, P, page 2.

1355. B. & H., *Abstracts*, 434.

1356. Book A, List of Applicants for Land in Austin's Colonies, 109.

1357. Spanish Archives, VIII, 975.

1358. Book B, Reports, P, page 8.

1359. B. & H., *Abstracts*, 434.

1360. Certificate of Character, Spanish Archives, issued to Dickinson Parker by Radford Berry, May 18, 1835.

1361. Book A, Reports, P, page 2.

1362. B. & H., *Abstracts*, 433.

1363. Book A, Reports, P, page 5.

1364. B. & H., *Abstracts*, 734.

1365. B. & H., *Abstracts*, 404, 433.

1366. Book A, Reports, P, page 1.

1367. B. & H., *Abstracts*, 433.

1368. Spanish Archives, XXIV, 295.

1369. B. & H., *Abstracts*, 433.

1370. Book A, Reports, N, page 2.

1371. B. & H., *Abstracts*, 431, 434.

1372. Book A, Reports, page 6.

1373. B. & H., *Abstracts*, 427.

1374. William H. Patton to S. F. Austin, November 2, 1832.

1375. Spanish Archives, XXIV, 311.

1376. B. & H., *Abstracts*, 434.

1377. Brown, *History of Texas*, II, 131.

1378. Book B, Reports, P, page 2.

1379. Muster Rolls, 28.

1380. Book B, Reports, P, page 2.

1381. B. & H., *Abstracts*, 433.

1382. Book B, Reports, P, page 2.

1383. B. & H., *Abstracts*, 433.

1384. Book A, Lists of Applicants for Land in Austin's Colonies, 79.

1385. Book A, Reports, P, page 2.

1386. B. & H., *Abstracts*, 433.

1387. Book A, Reports, P, page 1.

1388. B. & H., *Abstracts*, 433.

1389. Book A, List of Applicants for Land in Austin's Colonies, 75.

1390. Book B, Reports, P, page 1.

1391. B. & H., *Abstracts*, 431.

1392. B. & H., *Abstracts*, 411, 431.

1393. Book A, Reports, P, page 6.

1394. Spanish Archives, IV, 802.

1395. Book A, Reports, P, page 6.

1396. B. & H., *Abstracts*, 433.

1397. Book B, Reports, P, page 7.

1398. Spanish Archives, XIV, 201.

1399. Book B, Reports, P, page 7.

1400. B. & H., *Abstracts*, 433.

1401. B. & H., *Abstracts*, 407.

1402. Spanish Archives, XXXIV.

1403. B. & H., *Abstracts*, 433.

1404. Book A, Reports, P, page 7.

1405. Spanish Archives, V, 1450.

1406. Spanish Archives, V, 1450.

1407. B. & H., *Abstracts*, 433.

1408. Book B, List of Applicants for Land in Austin's Colonies, 97.

1409. Book A, Reports, P, page 3.

1410. B. & H., *Abstracts*, 434, 428.

1411. Certificate of Character, Spanish Archives, issued by John Bevil, Alcalde of Nacogdoches to Michael Pevetoe December 10, 1834.

1412. Book B, Reports, P, page 1.

1413. B. & H., *Abstracts*, 434, 431.

1414. Certificate of Character issued by Radford Berry to Samuel Philips, October 2, 1835

1415. Muster Rolls, 125.

1416. Book B, Reports, P, page 7.

1417. B. & H., *Abstracts*, 430.

1418. Book A, Reports, P, page 5.

1419. B. & H., *Abstracts*, 433.

1420. Book B, Reports, P, page 7.

1421. B. & H., *Abstracts*, 433.

1422. Book B, Reports, P, page 4.

1423. B. & H., *Abstracts*, 434, 429.

1424. B. & H., *Abstracts*, 409.

1425. Muster Rolls, 28.

1426. Book A, Reports, P, page 3.

1427. B. & H., *Abstracts*, 434, 428.

1428. Book A, Reports, P, page 4.

1429. B. & H., *Abstracts*, 434, 429.

1430. Book A, Reports, P, page 3.

1431. B. & H., *Abstracts*, 428, 433.

1432. List of Applicants for Land in Austin's Colonies, 99.

1433. Spanish Archives, XI, 645.

1434. Book B, Reports, P, page 1.

1435. B. & H., *Abstracts*, 434.

1436. Book B, Reports, P, page 4.

1437. B. & H., *Abstracts*, 433.

1438. B. & H., *Abstracts*, 416, 433.

1439. Book A, Reports, P, page 5.

1440. B. & H., *Abstracts*, 432, 434.

1441. Book A, Reports, P, page 3.

1442. B. & H., *Abstracts*, 433.

1443. Book B, Reports, R, page 1.

1444. Spanish Archives, II, 572.

1445. B. & H., *Abstracts*, 466.

1446. Book A, Reports, R, page 3.

1447. B. & H., *Abstracts*, 466.

1448. Spanish Archives, XIV, 336.

1449. B. & H., *Abstracts*, 461.

1450. Certificate of Character issued by Nathan Davis to Samuel Raymond, July 21, 1835. Spanish Archives.

1451. B. & H., *Abstracts*, 465.

1452. Book A, List of Applicants for Land in Austin's Colonies, 101.

1453. Book A, Reports, R, page 1.

1454. B. & H., *Abstracts*, 465.

1455. Book A, Reports, R, page 4.

1456. B. & H., *Abstracts*, 465.

1457. Book A, List of Applicants for Land in Austin's Colonies, 101.

1458. Book A, Reports, R, page 1.

1459. B. & H., *Abstracts*, 466.

1460. B. & H., *Abstracts*, 439, 463.

1461. Brown, *History of Texas*, II, 187.

1462. Book A, Reports, R, page 4.

1463. Spanish Archives, I, 349.

1464. Book A, Reports, R, page 4.

1465. B. & H., *Abstracts*, 465.

1466. Book A, Reports, R, page 4.

1467. Spanish Archives, X, 269.

1468. B. & H., *Abstracts*, 465.

1469. Book B, Reports, R, page 1.

1470. Book A, List of Applicants for Land in Austin's Colonies

1471. Spanish Archives, VIII, 511.

1472. Book B, Reports, R, page 1.

1473. B. & H., *Abstracts*, 465.

1474. T.J. Green, *Mier*, 441.

1475. Book B, Reports, R, page 1.

1476. B. & H., *Abstracts*, 465.

1477. Book B, Reports, R, page 5.

1478. Book A, Reports, R, page 4.

1479. B. & H., *Abstracts*, 465.

1480. B. & H., *Abstracts*, 442, 640.

1481. Book A, Reports, R, page 5.

1482. B. & H., *Abstracts*, 465.

1483. Book A, Reports, R, page 4.

1484. B. & H., *Abstracts*, 465.

1485. Certificate of Character issued to William Richardson by Radford Berry, September 17, 1835.

Spanish Archives.

1486. Spanish Archives, XXI, 1461.

1487. Book B, Reports, R, page 1.

1488. Book A, Reports, R, page 1.

1489. B. & H., *Abstracts*, 459.

1490. List of Applicants for Land in Wavell's Colony, 36.

1491. Book A, Reports, R, page 5.

1492. B. & H., *Abstracts*, 466.

1493. Thomas Robbins to G. A. Nixon, Spanish Archives.

1494. B. & H., *Abstracts*, 466.

1495. B. & H., *Abstracts*, 440, 466.

1496. Book B, Reports, U, page 1.

1497. George W. Robinson to Stephen F. Austin, June 15, 1834. Spanish Archives.

1498. Book B, Reports, U, page 3.

1499. B. & H., *Abstracts*, 465.

1500. Thrall, *Pictorial History*, 605.

1501. Robinson, "Recollections of Joel W. Robinson", in the *Quarterly of the Texas State Historical Association*, VI, 241-247.

1502. B. & H., *Abstracts*, 466.

1503. Book B, List of Applicants for Land in Austin's Colonies, 73.

1504. Book A, Reports, R, page 6.

1505. Book B, Reports, U, page 4.

1506. Book A, Reports, R, page 3.

1507. B. & H., *Abstracts*, 465.

1508. Thrall, *Pictorial History*, 606.

1509. B. & H., *Abstracts*, 439, 466.

1510. Brown, *History*, 334, 104.

1511. Thrall, *Pictorial History*, 606.

1512. Book B, Reports, R, page 5.

1513. Muster Rolls, 130.

1514. Book B, Reports, R, page 5.

1515. B. & H., *Abstracts*, 464.

1516. Certificate of Character issued by Vital Flore to James Rowe, December 3, 1834. Spanish Archives.

1517. Book B, Reports, R, page 2.

1518. B. & H., *Abstracts*, 465.

1519. B. & H., *Abstracts*, 441, 463.

1520. Book A, Reports, R, page 5.

1521. Spanish Archives, XXXV, 20.

1522. Book A, Reports, R, page 5.

1523. B. & H., *Abstracts*

1524. Thrall, *Pictorial History*, 607-608.

1525. Spanish Archives, XX, 55.

1526. Book B, Reports, R, page 1.

1527. B. & H., *Abstracts*, 462.

1528. B. & H., *Abstracts*, 464.

1529. Certificate of Character issued to John Sadler by Benjamin Lindsey, September 25, 1834.

1530. Spanish Archives, XX, 351.

1531. Book A, Reports, S, page 3.

1532. B. & H., *Abstracts*, 515.

1533. Book A, Reports, 51.

1534. B. & H., *Abstracts*, 515.

1535. Spanish Archives, XV, 1067.

1536. Book A, Reports, S, page 1.

1537. B. & H., *Abstracts*, 516.

1538. Certificate of Character issued to John Saunder by Judge of San Augustine, September 17, 1835.

1539. Muster Rolls, 133.

1540. Book A, Reports, S, page 4.

1541. B. & H., *Abstracts*, 133.

1542. Book B, List of Applicants for Land in Austin's Colonies, 93.

1543. B. & H., *Abstracts*, 516.

1544. Book A, Reports, S, page 3.

1545. B. & H., *Abstracts*, 514.

1546. Book A, Reports, S, page 4.

1547. B. & H., *Abstracts*, 516.

1548. Folson, *Mexico in 1842*, page 249.

1549. Book A, Reports, S, page 4.

1550. B. & H., *Abstracts*, 515.

1551. B. & H., *Abstracts*, 512.

1552. Book B, Reports, S, page 6.

1553. B. & H., *Abstracts*, 516.

1554. Book B, Reports, S, page 6.

1555. B. & H., *Abstracts*, 515.

1556. Brown, *History of Texas* II, 111.

1557. Book B, Reports, S, page 15.

1558. B. & H., *Abstracts*, 517.

1559. Book B, Reports, S, page 3.

1560. B. & H., *Abstracts*, 515.

1561. Thrall, *Pictorial History*, 616.

1562. B. & H., *Abstracts*, 516.

1563. B. & H., *Abstracts*, 467, 514, 505.

1564. B. & H., *Abstracts*, 515.

1565. Book A, Reports, S, page 6.

1566. B. & H., *Abstracts*, 515.

1567. Book B, Reports, S, page 1.

1568. B. & H., *Abstracts*, 516.

1569. B. & H., *Abstracts*, 468, 515, 508.

1570. Spanish Archives, XIII, 743.

1571. Book A, Reports, S, page 1.

1572. B. & H., *Abstracts*, 55.

1573. Brown, *History of Texas*, II, 141.

1574. C. W. Raines, *Bibliography*.

1575. Book A, Reports, S, page 7.

1576. B. & H., *Abstracts*, 515.

1577. C. W. Raines, *Bibliography*.

1578. Book B, List of Applicants for Land in Austin's Colonies, 59.

1579. Book A, Reports, S, page 7.

1580. B. & H., *Abstracts*, 512.

1581. Brown, *History of Texas*, II, 142.

1582. Book B, List of Applicants for Land in Austin's Colonies, 5.

1583. Spanish Archives, XII, 155.

1584. Book B, Reports, S, page 7.

1585. B. & H., *Abstracts*, 517.

1586. B. & H., *Abstracts*, 470, 506.

1587. Book A, Reports, S, page 6.

1588. B. & H., *Abstracts*, 510.

1589. Book A, Reports, S, page 4.

1590. Spanish Archives, XXX, 46.

1591. B. & H., *Abstracts*, 517.

1592. Book A, Reports, S, page 1.

1593. B. & H., *Abstracts*, 517.

1594. Book A, Reports, S, page 3.

1595. Thrall, *Pictorial History*, 619.

1596. Book A, Reports, S, page 7.

1597. B. & H., *Abstracts*, 517.

1598. Thrall, *Pictorial History*, 620.

1599. Thrall, *Pictorial History*, 620.

1600. Spanish Archives, XV, 927.

1601. Book A, Reports, S, page 4.

1602. Thrall, *Pictorial History*, 621.

1603. Book B, Reports, S, page 9.

1604. Book A, Reports, S, page 8.

1605. B. & H., *Abstracts*, 515.

1606. Book A, Reports, S, page 3.

1607. Spanish Archives, XXII, 247.

1608. Book A, Reports, S, page 3.

1609. B. & H., *Abstracts*, 516.

1610. Book A, Reports, 77.

1611. Book B, Reports, S, page 9.

1612. B. & H., *Abstracts*, 514.

1613. Book A, Reports, S, page 8.

1614. Muster Rolls, 125.

1615. Spanish Archives, XVIII, 255

1616. Book A, Reports, S, page 8.

1617. B. & H., *Abstracts*, 517.

1618. Book A, Reports, S, page 7.

1619. B. & H., *Abstracts*, 507.

1620. Book A, List of Applicants for Land in Austin's Colonies, 77.

1621. Book B, Reports, S, page 1.

1622. B. & H., *Abstracts*, 516.

1623. Book B, Reports, S, page 1.

1624. Spanish Archives, XXXIV, 68.

1625. Book B, Reports, S, page 11.

1626. B. & H., *Abstracts*, 515.

1627. Book A, Reports, S, page 7.

1628. B. & H., *Abstracts*, 515.

1629. Book B, List of Applicants for Land in Austin's Colonies, 21.

1630. Book B, List of Applicants for Land in Austin's Colonies, 21.

1631. Book A, Reports, S, page 1.

1632. B. & H., *Abstracts*, 514.

1633. Book A, Reports, S, page 11.

1634. Book A, Reports, S, page 1.

1635. Spanish Archives, VI, 2083.

1636. Thrall, *Pictorial History*, 622.

1637. Book A, Reports, S, page 1.

1638. B. & H., *Abstracts*, 516.

1639. Thrall, *Pictorial History*, 622.

1640. Book A, Reports, S, page 1.

1641. B. & H., *Abstracts*, 515.

1642. Book A, Reports, S, page 8.

1643. B. & H., *Abstracts*, 516.

1644. B. & H., *Abstracts*, 473, 505.

1645. Book B, Reports, S, page 10.

1646. B. & H., *Abstracts*, 514.

1647. Book A, Reports, S, page 6.

1648. B. & H., *Abstracts*, 514.

1649. Book A, Reports, S, page 10.

1650. B. & H., *Abstracts*, 515.

1651. Book B, Reports, S, page 9.

1652. B. & H., *Abstracts*, 514.

1653. Book A, List of Applicants for Land in Austin's Colonies, 33.

1654. B. & H., *Abstracts*, 515.

1655. Book A, List of Applicants for Land in Austin's Colonies, 89.

1656. Spanish Archives, VI, 1894.

1657. Wooten (Editor), *A Comprehensive History of Texas*, II, 904.

1658. B. & H., *Abstracts*, 516.

1659. Book A, List of Applicants for Land in Austin's Colonies, 33.

1660. Book A, Reports, H, page 4.

1661. Book A, List of Applicantsfor Land in Austin's Colonies, 33.

1662. Spanish Archives, VI, 1822.

1663. Book A, Reports, H, page 4.

1664. B. & H., *Abstracts*, 515.

1665. Book B, Reports, S, page 10.

1666. Spanish Archives, V, 1478.

1667. Book A, Reports, S, page 2.

1668. B. & H., *Abstracts*, 504, 515.

1669. B. & H., *Abstracts*, 504.

1670. B. & H., *Abstracts*, 516.

1671. Book A, Reports, S, page 3.

1672. B. & H., *Abstracts*, 515.

1673. Folson, *Mexico in 1842*, 249.

1674. Book B, Reports, S, page 1.

1675. Muster Rolls, 133.

1676. B. & H., *Abstracts*, 509.

1677. Book A, Reports, S, page 1.

1678. Book A, Reports, S, page 6.

1679. B. & H., *Abstracts*, 514.

1680. Thrall, *Pictorial History*, 625.

1681. Spanish Archives, VII, 241.

1682. Thrall, *Pictorial History*, 625.

1683. Book A, Reports, S, page 4.

1684. B. & H., *Abstracts*, 515.

1685. Thrall, *Pictorial History*, 625.

1686. Brown, *History of Texas*, II, 215.

1687. Thrall, *Pictorial History*, 625.

1688. W. T. Swain to Stephen F. Austin, May 23, 1834.

1689. Spanish Archives, XIV, 244.

1690. B. & H., *Abstracts*, 515.

1691. Book A, Reports, S, page 1.

1692. B. & H., *Abstracts*, 515.

1693. Book A, Reports, I, page 2.

1694. B. & H., *Abstracts*, 515.

1695. Spanish Archives,XI, 597.

1696. B. & H., *Abstracts*, 516.

1697. Book A, Reports, S, page 5.

1698. B. & H., *Abstracts*, 516.

1699. Book A, Reports, S, page 5.

1700. B. & H., *Abstracts*, 516.

1701. Brown, *History of Texas* II, 348.

1702. Brown, *History of Texas* II, 41.

1703. Book A, Reports, S, page 4.

1704. B. & H., *Abstracts*, 514.

1705. Book A, Reports, T, page 1.

1706. B. & H., *Abstracts*, 537.

1707. Book A, Reports, T, page 3.

1708. B. & H., *Abstracts*, 538.

1709. James E. Winston, "Kentucky and the Independence of Texas", *Southwestern Historical Quarterly*, XVI, 33.

1710. Book A, Reports, T, page 1.

1711. Muster Rolls, 28.

1712. Book A, Reports, T, page 1.

1713. B. & H., *Abstracts*, 537.

1714. List of Applicants for Land in Austin's Colonies, 59.

1715. Book A, Reports, T, page 3.

1716. Spanish Archives, XIII, 447.

1717. B. & H., *Abstracts*, 537.

1718. Muster Rolls, 115.

1719. Book A, Reports, T, page 1.

1720. B. & H., *Abstracts*, 537.

1721. Book B, Reports, T, page 1.

1722. B. & H., *Abstracts*, 537.

1723. Certificate of Character issued to Cyrus W. Thompson by Wm. Hardin, June 3, 1835.

1724. Book B, Reports, T, page 3.

1725. B. & H., *Abstracts*, 538.

1726. Book A, Reports, T, page 1.

1727. Muster Rolls, 44.

1728. Spanish Archives, II, 330.

1729. Book A, Reports, T, page 1.

1730. B. & H., *Abstracts*, 538.

1731. List of Applicants for Land in Austin's Colonies, 67.

1732. Spanish Archives, VI, 1730.

1733. Book B, Reports, T, page 2.

1734. B. & H., *Abstracts*, 538.

1735. Book A, Reports, T, page 3.

1736. Muster Rolls, 44.

1737. Book A, Reports, T, page 5.

1738. B. & H., *Abstracts*, 760.

1739. Book A, Reports, T, page 3.

1740. B. & H., *Abstracts*, 537.

1741. Book A, Reports, T, page 2.

1742. B. & H., *Abstracts*, 663.

1743. Book A, Reports, T, page 2.

1744. Book B, Reports, T, page 2.

1745. B. & H., *Abstracts*, 538.

1746. Book A, Reports, T, page 5.

1747. B. & H., *Abstracts*, 507.

1748. Certificate of Character issued to John B. Trenary by Wm. Whitley, September 5, 1835.

1749. B. & H., *Abstracts*, 538.

1750. List of Applicants for Land in Austin's Colonies, 99.

1751. Book A, Reports, T, page 5.

1752. B. & H., *Abstracts*, 538.

1753. Book A, Reports, T, page 3.

1754. B. & H., *Abstracts*, page 3.

1755. Thrall, *Pictorial History*, 629.

1756. Book B, Reports, T, page 1.

1757. B. & H., *Abstracts*, 538.

1758. Thrall, *Pictorial History*, 629.

1759. Book B, Reports, T, page 2.

1760. B. & H., *Abstracts*, 537.

1761. T. J. Green, *Mier*, 1842, 76.

1762. Book A, Reports, U, page 1.

1763. B. & H., *Abstracts*, 540.

1764. Brown, *History of Texas*, II, 111.

1765. Brown, *History of Texas*, II,186.

1766. Green, *Mier Expedition*, p. 76.

1767. Book B, Reports, V, page 1.

1768. Muster Rolls, 133.

1769. B. & H., *Abstracts*, 547.

1770. List of Applicants for Land in Milam's Colony, 43.

1771. B. & H., *Abstracts*, 347.

1772. Book A, Reports, V, page 1.

1773 Book A, Reports, V, page 2.

1774. Book A, Reports, V, page 2.

1775. Muster Rolls, 28.

1776. B. & H., *Abstracts*, 547.

1777. Book B, Reports, V, page 1.

1778. Book B, List of Applicants for Land in Austin's Colonies, 55.

1779. B. & H., *Abstracts*, 547.

1780. Book A, Reports, V, page 1.

1781. Spanish Archives, VIII, 779.

1782. Book B, Reports, H, page 1.

1783. B. & H., *Abstracts*, 582.

1784. Book B, List of Applicants for Land in Austin's Colonies, 79.

1785. Spanish Archives, I, 204.

1786. Book A, Reports, H, page 7.

1787. B. & H., *Abstracts*, 389.

1788. Book A, Reports, W, page 5.

1789. B. & H., *Abstracts*, 585.

1790. Book A, Reports, W, page 8.

1791. B. & H., *Abstracts*, 590.

1792. Book A, Reports, H, page 6.

1793. Muster Rolls, 126.

1794. Book A, Reports, H, page 6.

1795. B. & H., *Abstracts*, 591.

1796. Book B, Reports, W, page 10.

1797. B. & H., *Abstracts*, W, page 10.

1798. Book A, Reports, H, page 1.

1799. B. & H., *Abstracts*, 586.

1800. Book A, Reports, H, page 7.

1801. B. & H., *Abstracts*, 591.

1802. Book A, Reports, W, page 8.

1803. B. & H., *Abstracts*, 592.

1804. Book B, List of Applicants for Land in Austin's Colonies, 97.

1805. Book B, Reports, H, page 3.

1806. B. & H., *Abstracts*, 590.

1807. Certificate of Character issued to J. E. Watkins by Radford Berry, September 14, 1835.

1808. B. & H., *Abstracts*, 588.

1809. Book B, Reports, H, page 7.

1810. B. & H., Reports, 585.

1811. B. & H., *Abstracts*, 592.

1812. Book A, Reports, H, page 2.

1813. B. & H., *Abstracts*, 590.

1814. Book A, Reports, H, page 4.

1815. B. & H., *Abstracts*, 590.

1816. Muster Rolls, 115.

1817. B. & H., *Abstracts*, 550, 591.

1818. Book B, List of Applicants for Land by Austin's Colonists, 65.

1819. Book A, Reports, H, page 7.

1820. B. & H., *Abstracts*, 593.

1821. Book A, Reports, H, page 5.

1822. B. & H., *Abstracts*, 590.

1823. Brown, *History of Texas*, II, 187.

1824. Book A, Reports, H, page 1.

1825. B. & H., *Abstracts*, 590.

1826. Book A, Reports, H, page 8.

1827. B. & H., *Abstracts*, 592.

1828. Book B, Reports, H, page 4.

1829. B. & H., *Abstracts*, 588.

1830. Book A, Reports, H, page 6.

1831. B. & H., *Abstracts*, 589.

1832. Thrall, *Pictorial History*, 630.

1833. Book A, Reports, H, page 1.

1834. B. & H., *Abstracts*, 640.

1835. Brown, History, II, 88, 104.

1836. Book A, Reports, W, page 8.

1837. B. & H., *Abstracts*, 591.

1838. Book A, Reports, H, page 6.

1839. B. & H., *Abstracts*, 591.

1840. Book A, Reports, H, page 2.

1841. B. & H., *Abstracts*, 590.

1842. Book A, Reports, W, page 8.

1843. B. & H., *Abstracts*, 592.

1844. B. & H., *Abstracts*, 551.

1845. Book A, List of Applicants for Land in Austin's Colonies, 61.

1846. Book A, Reports, H, page 8.

1847. B. & H., *Abstracts*, 592.

1848. Book B, Reports, H, page 5.

1849. B. & H., *Abstracts*, 590.

1850. Book A, Reports, H, page 2.

1851. B. & H., *Abstracts*, 592.

1852. Book A, Reports, W, page 7.

1853. Book A, Reports, W, page 7.

1854. Book B, Reports, W, page 1.

1855. B. & H., *Abstracts*, 592.

1856. List of Applicants for Land in Robinson's Colony, 57.

1857. Book A, Reports, W, page 3.

1858. B. & H., *Abstracts*, 590.

1859. Book A, List of Applicants for Land in Austin's Colonies, 35.

1860. Book A, Reports, W, page 6.

1861. B. & H., *Abstracts*, W, page 6.

1862. Book B, List of Applicants for Land in Austin's Colonies, 47.

1863. Book B, Reports, W, page 3.

1864. Certificate of Character issued to Charles Williams by Benjamin Lindsey, October 27, 1834.

1865. Spanish Archives, XXIII, 1481.

1866. Book A, Reports, H, page 2.

1867. B. & H., *Abstracts*, 550.

1868. Certificate of Character issued to H. R. Williams by Benjamin Lindsey, October 27, 1834.

1869. Book A, Reports, H, page 2.

1870. B. & H., *Abstracts*, 554, 594.

1871. Book A, List of Applicants for Land in Austin's Colonies, 3.

1872. Book A, Reports, W, page 4.

1873. Book A, Reports, W, page 1.

1874. B. & H., *Abstracts*, 584.

1875. B. & H., *Abstracts*, 587.

1876. Book B, Reports, W, page 3.

1877. B. & H., *Abstracts*, 587.

1878. Muster Rolls, 44.

1879. Book A, Reports, W, page 1.

1880. B. & H., *Abstracts*, 582.

1881. T. J. Green, *Mier Expedition*, 443.

1882. Certificate of Character issued to Thomas Wilson by John Bodine, May 9, 1835.

1883. Spanish Archives, XXI, 1379.

1884. B. & H., *Abstracts*, 587.

1885. List of Applicants for Land in Milam's Colony, 9.

1886. Spanish Archives, XVI, 89.

1887. B. & H., *Abstracts*, 590.

1888. Book B, List of Applicants for Land in Austin's Colonies, 17.

1889. Book B, List of Applicants for Land in Austin's Colonies, 17.

1890. B. & H., *Abstracts*, 590.

1891. Book B, List of Applicants for Land in Austin's Colonies, 99.

1892. B. & H., *Abstracts*, 590.

1893. List of Orders of Survey, by Charles Taylor, Spanish Archives.

1894. B. & H., *Abstracts*, 590.

1895. James Washington Winters, "An Account of the Battle of San Jacinto", in the *Quarterly of the Texas State Historical Association*, VI, 139.

1896. B. & H., *Abstracts*, 590.

1897. Book A, Reports, W, page 7.

1898. Muster Rolls, 44.

1899. Spanish Archives, XXI, 1247.

1900. Book A, Reports, W, page 7.

1901. B. & H., *Abstracts*, 591.

1902. B. & H., *Abstracts*, 549, 590.

1903. B. & H., *Abstracts*, 558, 591.

1904. Muster Rolls, 115.

1905. B. & H., *Abstracts*, 587.

1906. Book B, List of Applicants for Land in Austin's Colonies, 9.

1907. Spanish Archives, X, 349.

1908. B. & H., *Abstracts*, 591.

1909. Book B, Reports, W, page 3.

1910. Book A, Reports, W, page 3.

1911. B. & H., *Abstracts*, 591.

1912. Book B, List of Applicants for Land in Austin's Colonies, 19.

1913. B. & H., *Abstracts*, 591.

1914. Spanish Archives, XXI, 927.

1915. B. & H., *Abstracts*, 592.

1916. Order for Survey of John Yancy's Land by George W. Smith, August 3, 1835, Spanish Archives.

1917. Muster Rolls, 126.

1918. Book B, Reports, Y, page 1.

1919. B. & H., *Abstracts*, 598.

1920. Certificate of Character issued to Swanson Yarbrough by Nathan Davis, August 18, 1835, Spanish Archives.

1921. Muster Rolls, 126.

1922. B. & H., *Abstracts*, 598.

1923. Book B, Reports, Y, page 1.

1924. Spanish Archives, XII, 815.

1925. B. & H., *Abstracts*, 598.

1926. Book B, Reports, Y, page 1.

1927. Book A, Reports, Y, page 1.

1928. B. & H., *Abstracts*, 598.

Bibliography

1. Sources.
    a. Manuscripts in the General Land Office.
        Muster Rolls.
        Books A and B, Lists of Applicants for Land in Austin's Colonies, Spanish Archives
        Lists of Applicants for Land in Milam's Colony, Spanish Archives.
        Lists of Applicants for Land in Wavell's Colony, Spanish Archives.
        Lists of Applicants for Land in Robertson's Colony, Spanish Archives.
        Letter Files, Spanish Archives.
        Index to Mexican and Spanish Titles, Spanish Archives.
        Sixty-three Volumes of Original Applications and Land Titles, Spanish Archives.
        Books A and B, Reports from the Land Boards.
2. Secondary Authorities.
        John Henry Brown, *History of Texas* II, St. Louis.
        Burlage and Hollingsworth, *Abstracts of Land Titles*, Austin, Texas, 1859.
        Charles J. Folsom, *Mexico in 1842*, New York, 1842.
        T. J. Green, *Journal of the Texian Expedition Against Mier*, New York, 1845.
        McMaster, *History of the People of the United States*, VI, New York, 1907.
        Thrall, *Pictorial History of Texas*, St. Louis, 1881.
        Herman von Holst, *Constitutional and Political History of the Unite States*, Freiburg, Germany, 1875
        Wooten (editor), *A Comprehensive History of the United States*, I, Dallas, 1898.
        H. Yoakum, *History of Texas*, II, New York, 1856.
        C. W. Raines, *Bibliography of Texas*, Austin, Texas,   1896.
        James Washington Winters, "An Account of the Battle of San Jacinto", in the *Quarterly of the Texas State Historical Association* VI, 139-145, Austin, 1903.
        Robinson, "Recollections of Joel W. Robinson", in the *Quarterly of the Texas State Historical Association* VI, 241-247, Austin, 1903.

James E. Winston, "Kentucky and the Independence of Texas", in the *Southwestern Historical Quarterly* XVI, 27-63, Austin, 1913.

James E. Winston, "Virginia and the Independence of Texas", *Southwestern Historical Quarterly* XVI, 277-284, Austin, 1913.